# JIG AND
# FIXTURE
# HANDBOOK

# JIG AND FIXTURE HANDBOOK

**CARR LANE MFG. CO.**

Third Edition

# JIG AND FIXTURE HANDBOOK

Carr Lane Mfg. Co.
P.O. Box 191970, 4200 Carr Lane Ct.
St. Louis, Missouri 63119-2196
314/647-6200

Printed in the United States of America
ISBN-0-9622079-1-8
Third Edition

# TABLE OF CONTENTS

# 1

# INTRODUCTION TO WORKHOLDING

Over the past century, manufacturing has made considerable progress. New machine tools, high-performance cutting tools, and modern manufacturing processes enable today's industries to make parts faster and better than ever before. Although workholding methods have also advanced considerably, the ***basic principles*** of clamping and locating are still the same.

## HISTORY

The first manufactured products were made one at a time. Early artisans started with little more than raw materials and a rough idea of the finished product. They produced each product piece by piece, making each part individually and fitting the parts into the finished product. This process took time. Moreover, the quality and consistency of products varied from one artisan to the next. As they worked, early manufacturing pioneers realized the need for better methods and developed new ideas.

Eventually, they found the secret of mass production: standardized parts. Standard parts not only speeded production, they also ensured the interchangeability of parts. The idea may be obvious today, but in its time, it was revolutionary.

These standard parts were the key to enabling less-skilled workers to replicate the skill of the craftsman on a repetitive basis. The original method of achieving consistent part configuration was the template. Templates for layout, sawing, and filing permitted each worker to make parts to a standard design. While early templates were crude, they at least gave skilled workers a standard form to follow for the part. Building on the template idea, workers constructed other guides and workholders to make their jobs easier and the results more predictable. These guides and workholders were the ancestors of today's jigs and fixtures.

Yesterday's workholders had the same two basic functions as today's: securely holding and accurately locating a workpiece. Early jigs and fixtures may have lacked modern refinements, but they followed many of the same principles as today's workholder designs.

## DEFINITIONS

Often the terms "jig" and "fixture" are confused or used interchangeably; however, there are clear distinctions between these two tools. Although many people have their own definitions for a jig or fixture, there is one universal distinction between the two. Both jigs and fixtures hold, support, and locate the workpiece. A jig, however, *guides* the cutting tool. A fixture *references* the cutting tool. The differentiation between these types of workholders is in their relation to the cutting tool. As shown in Figure 1-1, jigs use drill bushings to support and guide the tool. Fixtures, Figure 1-2, use set blocks and thickness, or feeler, gages to locate the tool relative to the workpiece.

### Jigs

The most-common jigs are drill and boring jigs. These tools are fundamentally the same. The difference lies in the size, type, and placement of the drill bushings. Boring jigs usually have larger bushings. These bushings may also have internal oil grooves to keep the boring bar lubricated. Often, boring jigs use more than one bushing to support the boring bar throughout the machining cycle.

In the shop, drill jigs are the most-widely used form of jig. Drill jigs are used for drilling, tapping, reaming, chamfering, counterboring, countersinking, and similar operations. Occasionally, drill jigs are used

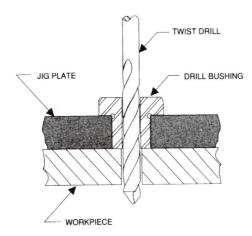

**Figure 1-1.** *A jig guides the cutting tool, in this case with a bushing.*

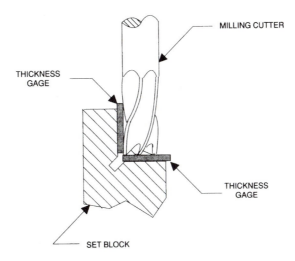

**Figure 1-2.** *A fixture references the cutting tool, in this case with a set block.*

to perform assembly work also. In these situations, the bushings guide pins, dowels, or other assembly elements.

Jigs are further identified by their basic construction. The two common forms of jigs are open and closed. Open jigs carry out operations on only one, or sometimes two, sides of a workpiece. Closed jigs, on the other hand, operate on two or more sides. The most-common open jigs are template jigs, plate jigs, table jigs, sandwich jigs, and angle plate jigs. Typical examples of closed jigs include box jigs, channel jigs, and leaf jigs. Other forms of jigs rely more on the application of the tool than on their construction for their identity. These include indexing jigs, trunnion jigs, and multi-station jigs.

Specialized industry applications have led to the development of specialized drill jigs. For example, the need to drill precisely located rivet holes in aircraft fuselages and wings led to the design of large jigs, with bushings and liners installed, contoured to the surface of the aircraft. A portable air-feed drill with a bushing attached to its nose is inserted through the liner in the jig and drilling is accomplished in each location.

## Fixtures

Fixtures have a much-wider scope of application than jigs. These workholders are designed for applications where the cutting tools cannot be guided as easily as a drill. With fixtures, an edge finder, center finder, or gage blocks position the cutter. Examples of the more-common fixtures include milling fixtures, lathe fixtures, sawing fixtures, and grinding fixtures. Moreover, a fixture can be used in almost any operation that requires a precise relationship in the position of a tool to a workpiece.

Fixtures are most often identified by the machine tool where they are used. Examples include mill fixtures or lathe fixtures. But the function of the fixture can also identify a fixture type. So can the basic construction of the tool. Thus, although a tool can be called simply a mill fixture, it could also be further defined as a straddle-milling, plate-type mill fixture. Moreover, a lathe fixture could also be defined as a radius-turning, angle-plate lathe fixture. The tool designer usually decides the specific identification of these tools.

## Tool or Tooling

The term "tool" encompasses both jigs and fixtures. Essentially, it is a generic term describing a workholder which is identified with a part or machine. Sometimes "tool" is used to refer to a cutting tool or a machine tool, so it is important to make clear distinctions.

## Workholders

Another term which describes both jigs and fixtures is "workholder." A broad term, it frequently identifies any device which holds, supports, and locates a workpiece. In addition to jigs and fixtures, vises, collets, clamps, and other similar devices are also workholders.

## PERMANENT AND TEMPORARY WORKHOLDERS

Jigs and fixtures are most often found where parts are produced in large quantities, or produced to complex specifications for a moderate quantity. With the same design principles and logic, workholding devices can be adapted for limited-production applications. The major difference between permanent and temporary workholders is the cost/benefit relationship between the workholder and the process. Some applications require jigs and fixtures solely for speed; others require less speed and higher precision. The requirements of the application have a direct impact on the type of jig or fixture built and, consequently, the cost.

## Permanent Jigs and Fixtures

Workholders for high-volume production are usually permanent tools. These permanent jigs and fixtures are most often intended for a single operation on one particular part. The increased complexity of permanent workholders yields benefits in improved productivity and reduced operator decision-making, which result in the tool having a lower average cost per unit or per run. Therefore, more time and money can be justified for these workholders.

In the case of hydraulic or pneumatic fixtures, inherent design advantages can dramatically improve productivity and, hence, reduce per-unit costs even further, even though the initial cost to construct these fixtures is the most expensive of all fixture alternatives. In some cases, where machine-loading considerations are paramount, such as a pallet-changing machining center, even duplicate permanent fixtures may be justified.

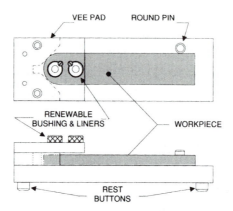

**Figure 1-3**. *A permanent workholder used for a drilling operation.*

Permanent jigs and fixtures are typically constructed from standard tooling components and custom-made parts. Figure 1-3 shows a typical permanent workholder for a drilling operation.

Low-volume runs and ones with fewer critical dimensions are often produced with throwaway jigs and fixtures. These tools would typically be one-time-use items constructed from basic materials at hand and discarded after production is complete. Although throwaway jigs and fixtures are technically permanent workholders, in effect they are actually temporary.

## General-Purpose Workholders

In many instances, the shape of the part and the machining to be performed allow for the use of a general-purpose workholder such as a vise, collet, or chuck. These workholders are adaptable to different machines and many different parts.

Since they are not part-specific, their versatility allows for repeated use on a variety of different or limited-production runs. The cost of these workholders would usually be averaged over years and might not even be a factor in job-cost calculations. The general-purpose nature of these workholders necessitates a higher level of operator care and attention to maintain consistency and accuracy. For these reasons, general-purpose workholders are not preferred for lengthy production runs.

## Modular Fixtures

Modular fixtures achieve many of the advantages of a permanent tool using only a temporary setup. Depicted in Figure 1-4, these workholders combine ideas and elements of permanent and general-purpose workholding.

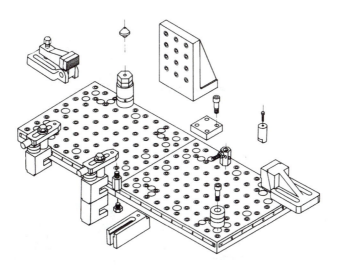

**Figure 1-4.** *Modular workholders combine ideas and elements of both permanent and temporary workholding to make inexpensive-yet-durable workholders.*

The primary advantage of modular fixtures is that a tool with the benefits of permanent tooling (setup reduction, durability, productivity improvements, and reduced operator decision-making) can be built from a set of standard components. The fixture can be disassembled when the run is complete, to allow the reuse of the components in a different fixture. At a later time the original can be readily reconstructed from drawings, instructions, and photographic records. This reuse enables the construction of a complex, high-precision tool without requiring the corresponding dedication of the fixture components.

Figure 1-5 shows how modular fixturing fits into the hierarchy of workholding options, ranking below permanent fixturing yet above general-purpose workholders. Virtually every manufacturer has good applications for each of these three options at one time or another.

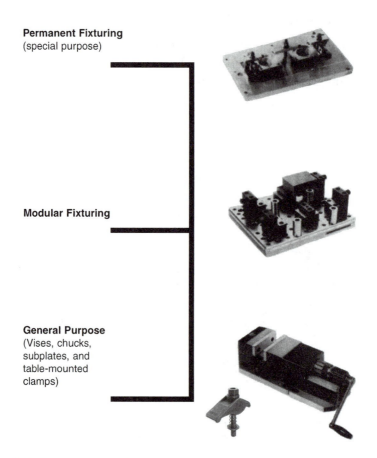

**Permanent Fixturing**
(special purpose)

**Modular Fixturing**

**General Purpose**
(Vises, chucks,
subplates, and
table-mounted
clamps)

**Figure 1-5.** *The hierarchy of workholding options.*

## DESIGN CONSIDERATIONS

The principal considerations when choosing among workholder varieties fall into three general categories: tooling cost, tooling details, and tooling operation. Although each of these categories is separated here, in practice they are interdependent. The following are some design differences and considerations for permanent, general-purpose, and modular workholders.

## Tooling Costs

The total cost of any jig or fixture is frequently the major area of consideration in many workholder designs. Although initial cost is a major element, it should not be the basis for accepting or rejecting any tooling option.

A more-proper economic evaluation of the workholder design takes into consideration many other factors. As discussed previously, permanent fixtures have distinct advantages in the production of high-volume and high-precision parts. They also typically reduce machine setup time, machine cycle time, and the level of operator skill required to produce satisfactory quality output. Over a long production run, or a series of runs in the life of a tool, the average cost of the tool per piece produced can be quite low.

General-purpose workholders are more expensive than temporary tools in most cases, but their utility and flexibility often allow these workholders to be regarded as a capital cost to be amortized over a period of time without regard to actual usage. Similarly, modular fixturing is typically a capital investment to be amortized over a set lifespan, with an average cost assigned to usage for each anticipated job.

Another cost to be considered is workholder disposition. Permanent fixtures require storage and maintenance to keep them available for their next use. General-purpose tools are reused extensively, but still incur some costs for maintenance and storage. Similarly, modular fixtures will be disassembled, and the components maintained, stored, and reused frequently.

## Tooling Details

Tooling details are the overall construction characteristics and special features incorporated into the jig or fixture. Permanent workholders are designed and built to last longer than temporary workholders. So, permanent jigs and fixtures usually contain more-elaborate parts and features than temporary workholders.

There are several other differences between permanent and temporary workholders in this area. These include the type and complexity of the individual tooling elements, the extent of secondary machining and finishing operations on the tool, the tool-design process, and the amount of detail in the workholder drawings. Since the elements for modular workholders are usually part of a complete set, or system, only rarely will additional custom components need to be made.

Permanent workholders contain different commercial tooling

components based on expected tool usage. Permanent jigs intended for a high-volume drilling operation, for example, often use a renewable bushing and liner bushing together. A throwaway jig for a smaller production run often uses a simple press-fit bushing.

The secondary operations normally associated with tooling include hardening, grinding, and similar operations to finish the workholder. Usually, permanent workholders are hardened and ground to assure their accuracy over a long production run. Since they are intended only for short production runs, throwaway jigs and fixtures do not require these operations. Another secondary operation frequently performed on permanent tools, but not temporary tools, is applying a protective finish, such as black oxide, chrome plating, or enamel paint.

In designing a permanent workholder, the designer often makes detailed engineering drawings to show the toolroom exactly what must be done to build the workholder. With temporary workholders, the design drawings are often sent to the toolroom as simple freehand sketches.

Permanent tools are normally designed for long-term use. This being the case, the drawings and engineering data for the permanent jig or fixture then become a permanent record. With modular workholders, the designer may either construct drawings or specify building the workholder directly around the part. Here only a parts list and photographs or video tape are kept as a permanent record.

Certain workholding applications require special fixture characteristics. For example, a particularly corrosive environment may require stainless steel components and clamps to deliver a satisfactory life cycle. In other cases, variable workpiece dimensions, as in a casting, necessitate clamping devices which can compensate for these variations. Appearance of a finished part might require the use of nylon, plastic, or rubber contact points to protect the part.

Similarly, the selection of tooling details can enhance the productivity of some permanent tools. For example, utilizing small hydraulic clamps may allow loading many parts on a workholder due to the compactness of the design. This would enhance productivity by reducing  load/unload time as a percentage of total cycle time. Duplicate fixtures are sometimes justified for machining centers because they allow loading of parts on one pallet during the machining cycle on the other pallet.

## Tooling Operation

The performance of any workholder is critical to the complete usefulness of the tool. If the workholder cannot perform the functions desired in the manner intended, it is completely useless, regardless of the cost or the extent of the detail. As the performance of a permanent, modular, or general-purpose workholder is considered, several factors about the machine tools must be known. These factors include the type, size, and number of machine tools needed for the intended operations.

Workholders are sometimes designed to serve multiple functions. For example, it is possible to have a workholder that acts both as a drill jig and a milling fixture. These tools are called combination tools or multiple-function workholders. Figure 1-6 shows a typical temporary workholder for drilling and milling operations on the same part. In this example, since the workholder has provisions for both milling and drilling, it is classified as both a drill jig and milling fixture.

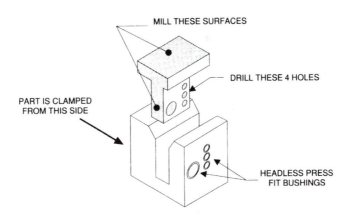

MILL THESE SURFACES

DRILL THESE 4 HOLES

PART IS CLAMPED
FROM THIS SIDE

HEADLESS PRESS
FIT BUSHINGS

**Figure 1-6.** *A combination drill jig/milling fixture used for both types of operations on the same part.*

Other machine considerations may come into play as well. On numerically controlled machines, for example, care must be taken in fixture design to position clamps out of the cutting tool's path. Pallet machines require different fixtures than other machines. Obviously, vertical mills would be tooled differently than horizontal mills. Likewise, the way parts are loaded onto the fixture has implications for fixture design.

**EXTERNAL-MACHINING APPLICATIONS:**
Flat-Surface Machining
  • Milling fixtures
  • Surface-grinding fixtures
  • Planing fixtures
  • Shaping fixtures

Cylindrical-Surface Machining
  • Lathe fixtures
  • Cylindrical-grinding fixtures

Irregular-Surface Machining
  • Band-sawing fixtures
  • External-broaching fixtures

**INTERNAL-MACHINING APPLICATIONS:**
Cylindrical- and Irregular-Hole Machining
  • Drill jigs
  • Boring jigs
  • Electrical-discharge-machining fixtures
  • Punching fixtures
  • Internal-broaching fixtures

**NON-MACHINING APPLICATIONS:**
Assembly
  • Welding fixtures
  • Mechanical-assembly fixtures
    (Riveting, stapling, stitching, pinning, etc.)
  • Soldering fixtures

Inspection
  • Mechanical-inspection fixtures
  • Optical-inspection fixtures
  • Electronic-inspection fixtures

Finishing
  • Painting fixtures
  • Plating fixtures
  • Polishing fixtures
  • Lapping fixtures
  • Honing fixtures

Miscellaneous
  • Layout templates
  • Testing fixtures
  • Heat-treating fixtures

**Figure 1-7.** *Typical applications of jigs and fixtures.*

Despite the workholder design or the size of the production run, every jig or fixture must meet certain criteria to be useful. These criteria include accuracy, durability, and safety. Accuracy, with regard to jigs and fixtures, is the ability of a workholder to produce the desired result, within the required limits and specifications, part after part, throughout the production run.

To perform to this minimum level of accuracy, the workholder must also be durable. So, the jig or fixture must be designed and built to maintain the required accuracy throughout the expected part production. If part production is continuous, year after year, the jig or fixture must be more durable than is necessary for only one production run.

The final consideration, safety, is actually the most important. No matter how good the design or construction, or how well it produces the desired accuracy, if the workholder is not safe, it is useless. Safety is a primary concern in the design of any workholder.

Safety, as well as speed and reliability of part loading, can often be improved by the use of power clamping, either pneumatic or hydraulic. Once set, power clamps will repeatedly clamp with the identical force. This is not always true with manual clamps, which depend on operator diligence for the proper application of clamping force. In addition, power-clamping systems can have interlocks to the machine control which will shut the machine down if the system loses power—a clear safety advantage for both operator and machine tool.

## APPLICATIONS FOR JIGS AND FIXTURES

Typically, the jigs and fixtures found in a machine shop are for machining operations. Other operations, however, such as assembly, inspection, testing, and layout, are also areas where workholding devices are well suited. Figure 1-7 shows a list of the more-common classifications and applications of jigs and fixtures used for manufacturing. There are many distinct variations within each general classification, and many workholders are actually combinations of two or more of the classifications shown.

# 2

# CREATIVE TOOL DESIGN

The first step in designing any jig or fixture is a thorough evaluation of its functional requirements. The goal is to find a balanced combination of design characteristics at a reasonable cost. The part itself, processing, tooling, and machine-tool availability may all affect the extent of planning needed. Preliminary analysis may take from a few hours up to several days for more-complicated designs.

To design a workholder, begin with a logical and systematic plan. With a complete analysis, very few design problems occur. Workholder problems occur when design requirements are forgotten or underestimated. No specific formula or method works for every design, but the designer can employ a deliberate and logical system in the initial planning and design.

Tool design is essentially an exercise in problem solving. Creative problem solving can be described as a five-step process: 1) Identifying and defining the problem; 2) Gathering and analyzing information; 3) Brainstorming for alternative solutions; 4) Choosing the best solution; 5) Implementing the solution. This five-step process adapted to jig-and-fixture designs shown in Figure 2-1.

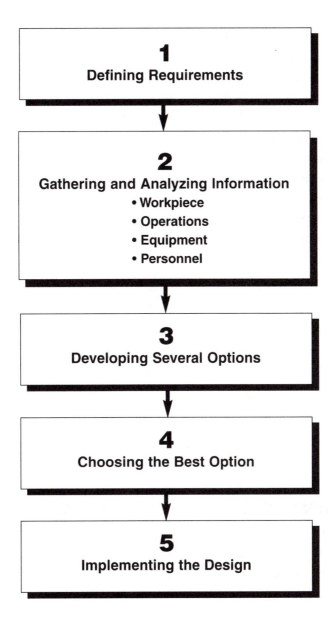

**Figure 2-1**. *Five steps that make up a good, systematic tool-design process.*

## DEFINING REQUIREMENTS

The first step in the tool-design process should be to clearly state the problem to be solved, or needs to be met. These requirements should be stated as broadly as possible, but specifically enough to define the scope of the design project.

The new tooling might be required either for first-time production of a new product, or to improve production of an existing part. When improving an existing job, the goal might be greater accuracy, faster cycle times, or both. Tooling might be designed for one part, or an entire family.

Tool design is an integral part of the product-planning process, interacting with product design, manufacturing, and marketing. To reach an optimum solution, all four of these groups need to work together concurrently.

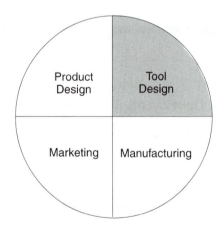

## GATHERING AND ANALYZING INFORMATION

In the second design phase, all data is collected and assembled for evaluation. The main sources of information are the part print, process sheets, and machine specifications. When collecting this information, make sure that part documents and records are current. For example, verify that the shop print is the current revision, and the processing information is up to date. Check with the product-design department for pending part revisions.

# CHECKLIST FOR DESIGN CONSIDERATIONS

**WORKPIECE:**
- ❏ Size (Large, Small)
- ❏ Shape (Rectangular, Square, Cylindrical, Spherical, Other)
- ❏ Required Accuracy (Tolerances, Machining allowances)
- ❏ Material Type (Steel, Stainless steel, Aluminum, Iron, Other)
- ❏ Material Condition (Cold rolled, Hot rolled, Cast, Forged, Other)
- ❏ Locating Points (Machined surfaces, Unmachined surfaces, Holes, Slots, Other)
- ❏ Locating Stability (Rigid, Fragile)
- ❏ Clamping Surfaces (Machined vs. unmachined, Supported vs. unsupported, Avoiding part damage)
- ❏ Production Quantity (Limited vs. mass production, One-time vs. recurring production, Product lifetime, Projected future increases)
- ❏ Pending part-design revisions

**OPERATIONS:**
- ❏ Types of Operations (Machining, Assembly, Other)
- ❏ Number of Separate Operations (Similar vs. different, Sequential vs. simultaneous)
- ❏ Sequence (Primary operations, Secondary operations, Heat treating, Finishing)
- ❏ Inspection Requirements

**EQUIPMENT:**
- ❏ Machine Tools (Horsepower, Size limitations, Weight limitations, Other)
- ❏ Cutting Tools
- ❏ Special Machinery
- ❏ Assembly Equipment and Tools
- ❏ Inspection Equipment and Tools
- ❏ Equipment Availability and Scheduling
- ❏ Plant Space Required

**PERSONNEL:**
- ❏ Safety Equipment (Machine, Operator, Plant)
- ❏ Safety Regulations and Work Rules
- ❏ Economy of Motion (Unloading, Loading, Clamping)
- ❏ Operator Fatigue
- ❏ Power Equipment Available
- ❏ Possible Automation

**Figure 2-2.** *Considerations when gathering and analyzing information for a tool design.*

An important part of the evaluation process is notetaking. Complete, accurate notes allow the designer to record important information. All ideas, thoughts, observations, and any other data about the part or tool are then available for later reference. It is always better to have too many ideas about a particular design than not enough. Good notes also minimize the chance that good ideas will be lost.

Four categories of design considerations need to be taken into account at this time: the workpiece, manufacturing operations, equipment, and personnel. A checklist is shown in Figure 2-2.

These categories, while separately covered here, are actually interdependent. Each is an integral part of the evaluation phase and must be thoroughly thought out before beginning the workholder design.

## Workpiece Considerations

Workpiece specifications are usually the most-important factors and have the largest influence on the workholder's final design. Typically these considerations include the size and shape of the part, the accuracy required, the properties of the part material, locating and clamping surfaces, and the number of pieces.

## Operation Considerations

These considerations include the type of operations required for the part, the number of operations performed, the sequence of operations, inspection requirements, and time restrictions.

## Equipment Considerations

Equipment considerations control the type of equipment needed to machine, assemble, and inspect a part. Often the available equipment determines whether the workholder is designed for single or multiple parts. A process engineer sometimes selects the equipment for required functions before the tool designer begins the design. Still, the tool designer should verify equipment choices for each operation.

A vertical milling machine, for example, is well suited for some drilling operations. But for operations that require a drill jig, a drill press is the most-cost-effective machine tool. Typically, equipment criteria include the following factors: types and sizes of machine tools, inspection equipment, scheduling, cutting tools, and general plant facilities.

## Personnel Considerations

Personnel considerations deal with the end user, or operator, of the equipment. Most special tools are designed to be used by shop personnel, so the design of any workholder must be made with the operator in mind. The first and most-important consideration in this phase is safety. No tool should ever be designed without complete safety in mind.

Additional factors typically considered in this category are operator fatigue, efficiency, economy of motion, and the speed of the operation. The designer must also know and understand the general aspects of design safety and all appropriate government and company safety rules and codes.

## DEVELOPING SEVERAL OPTIONS

The third phase of the tool-design process requires the most creativity. A typical workpiece can be located and clamped many different ways. An important strategy for successful tool design is brainstorming for several good tooling alternatives, not just choosing one path right away.

During this phase, the designer's goal should be adding options, not discarding them. In the interest of economy, alternative designs should only be developed far enough to make sure they are feasible, and to do a cost estimate.

## Brainstorming for Ideas

The designer usually starts with at least three options: permanent, modular, and general-purpose workholding, as seen in Figure 2-3. Each of these options has many clamping and locating options of its own. The more standard locating and clamping devices that a designer is familiar with, the more creative he can be.

There is seldom only one way to locate a part. Options include flat exterior surfaces (machined and unmachined), cylindrical and curved exterior surfaces, and internal features (such as holes and slots). The choice of standard locating devices is quite extensive.

Similarly, there are countless ways to clamp a part. For example, a workpiece can be clamped from the top, by gripping its outside edge,

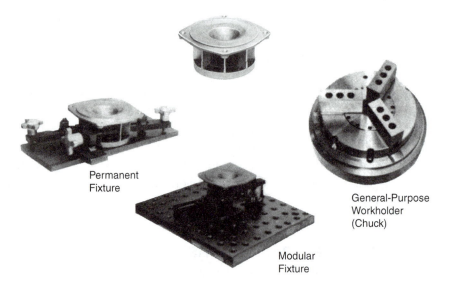

Permanent
Fixture

General-Purpose
Workholder
(Chuck)

Modular
Fixture

**Figure 2-3.** *Most tool-design projects begin with three general options: permanent, modular, and general-purpose workholding. Different methods of locating and clamping further increase the number of options.*

or gripping an internal surface. The choice of standard clamping devices is also very broad.

## Design Sketches

For preliminary sketches of the tool, one good idea is to use several colored pencils. Often, black is used to sketch the tool, red for the part, and blue for the machine tool. The different colors allow you to see, at a glance, which areas of the sketch show what part of the assembled unit. Another idea: use graph paper to keep the sketch proportional. Either plain or isometric graph paper works well for most design sketches.

The exact procedure used to construct the preliminary design sketches is not as important as the items sketched. For the most part, the preliminary sketch should start with the part to be fixtured. The required locating and supporting elements should be the next items added, including a base. The next step is to sketch the clamping devices. Once these elements are added to the sketch, the final items to add are the machine tool and cutters. Sketching these items together on the preliminary design sketch helps identify any problem areas in the design of the complete workholder.

## CHOOSING THE BEST OPTION

The fourth phase of the tool-design process is a cost/benefit analysis of different tooling options. Some benefits, such as greater operator comfort and safety, are difficult to express in dollars but are still important. Other factors, such as tooling durability, are difficult to estimate. Cost analysis is sometimes more of an art than a science.

Workholder-cost analysis compares one method to another, rather than finding exact costs. So, even though the values used must be accurate, estimates are acceptable. Sometimes these methods compare both proposed tools and existing tools, so, where possible, actual production data can be used instead of estimates.

### Initial Tooling Cost

The first step of evaluating the cost of any alternative is estimating the initial cost of the workholder. Add the cost of each element to the labor expense needed to design and build the jig or fixture.

To make this estimate, an accurate sketch of the tool is made first. Each part and component of the tool is numbered and listed individually. Here it is important to have an orderly method to outline this information. Figure 2-4 shows one way to make this listing. The exact appearance of the form is unimportant; only the information is important.

The next step is calculating the cost of material and labor for each tool element. Once again, it is important to have an orderly system of listing the data. First list the cost of each component, then itemize the operations needed to mount, machine, or assemble that component. Once these steps are listed, estimate the time required for each operation for each component, then multiply by the labor rate. This amount should then be added to the cost of the components and the cost of design to find the estimated cost of the workholder.

For modular fixtures, total component cost should be amortized over the system's typical lifetime. Although somewhat arbitrary, dividing total component cost by 100 (10 uses per year, for ten years) gives a fair estimate.

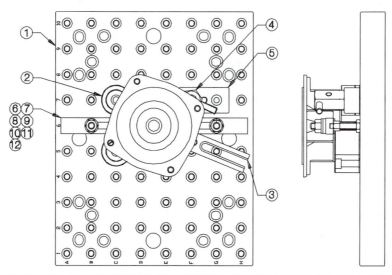

| ITEM | QTY. | PART NO. | DESCRIPTION | COMMENTS |
|------|------|----------|-------------|----------|
| 1 | 1 | CL-MF40-0101 | RECTANGULAR TOOLING PLATE | |
| 2 | 2 | CL-MF40-3204 | SHOULDER SUPPORT CYLINDER | |
| 3 | 1 | CL-MF40-3302 | EXTENSION SUPPORT | |
| 4 | 1 | CL-MF40-2901 | ADJUSTABLE STOP | |
| 5 | 1 | CL-MF40-3603 | HALFWAY EXTENSION SUPPORT | |
| 6 | 2 | CL-8-CS | CLAMP STRAP | |
| 7 | 2 | CL-4-CR | CLAMP REST | |
| 8 | 2 | CL-9-SPG | CLAMP SPRING | |
| 9 | 2 | CL-2-FW | FLAT WASHER | |
| 10 | 4 | CL-8-JN | JAM NUT | |
| 11 | 2 | CL-1/2-13X5.50 | STUD | |
| 12 | 2 | CL-3-SNW | SPHERICAL NUT AND WASHER | |

**Figure 2-4.** *Itemized listing of components for a workholder.*

## Cost Comparison

The total cost to manufacture a part is the sum of per-piece run cost, setup cost, and tooling cost. Expressed as a formula:

$$\frac{\text{Cost}}{\text{per Part}} = \text{Run Cost} + \frac{\text{Setup Cost}}{\text{Lot Size}} + \frac{\text{Initial Tooling Cost}}{\text{Total Qty over Tooling Lifetime}}$$

The following example shows three tooling options for the part in Figure 2-3: 1) a modular fixture; 2) a permanent fixture; 3) a permanent fixture using hydraulic power workholding. Each variable in the cost equation is explained separately below.

**Run Cost.** This is the variable cost per piece to produce a part, at shop labor rate (material cost does not need to be included as long as it is the same for all fixturing options). In our example, run costs for the permanent and modular fixtures are the same, while power workholding lowers costs by improving cycle time and reducing scrap.

|  | Modular Fixture | Permanent Fixture | Permanent Hydraulic Fixture |
|---|---|---|---|
| Run Cost | $4.50 | $4.50 | $3.50 |

**Setup Cost.** This is the cost to retrieve a fixture, set it up on the machine, and return it to storage after use. The permanent fixture is fastest to set up, the power-workholding fixture is slightly slower due to hydraulic connections, and the modular fixture is slowest due to the assembly required.

|  | Modular Fixture | Permanent Fixture | Permanent Hydraulic Fixture |
|---|---|---|---|
| Setup Cost | $240 | $80 | $100 |

**Lot Size.** This is the average quantity manufactured each time the fixture is set up. In our example, lot size is the same for all three options.

|  | Modular Fixture | Permanent Fixture | Permanent Hydraulic Fixture |
|---|---|---|---|
| Lot Size (Pieces) | 100 | 100 | 100 |

**Initial Tooling Cost.** This is the total cost of labor plus material to design and build a fixture (as explained in the previous section). The modular fixture is least expensive because components can be reused, the permanent fixture next, and the hydraulic fixture most expensive.

|  | Modular Fixture | Permanent Fixture | Permanent Hydraulic Fixture |
|---|---|---|---|
| Initial Tooling Cost | $341 | $1632 | $3350 |

*Total Quantity Over Tooling Lifetime.* This quantity, the last remaining variable, is the lesser of 1) total anticipated production quantity and 2) the quantity that can be produced before the tooling wears out. The following results are obtained by evaluating the cost-per-part formula at different lifetime quantities:

$$\frac{\text{Cost}}{\text{per}} = \frac{\text{Run}}{\text{Cost}} + \frac{\text{Setup Cost}}{\text{Lot Size}} + \frac{\text{Initial Tooling Cost}}{\text{Total Qty over Tooling Lifetime}}$$

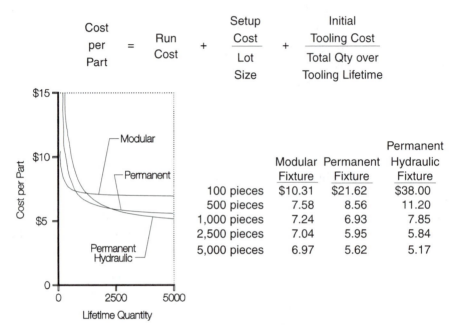

| | Modular Fixture | Permanent Fixture | Permanent Hydraulic Fixture |
|---|---|---|---|
| 100 pieces | $10.31 | $21.62 | $38.00 |
| 500 pieces | 7.58 | 8.56 | 11.20 |
| 1,000 pieces | 7.24 | 6.93 | 7.85 |
| 2,500 pieces | 7.04 | 5.95 | 5.84 |
| 5,000 pieces | 6.97 | 5.62 | 5.17 |

For a one-time run of 100 pieces, the modular fixture is clearly the most economical. If ten runs (1,000 pieces) are expected, the permanent fixture is best. For 2,500 pieces and above, the power-workholding fixture would be the best choice. This analysis assumes that all noneconomic factors are equal.

## IMPLEMENTING THE DESIGN

The final phase of the tool-design process consists of turning the chosen design approach into reality. Final details are decided, final drawings are made, and the tooling is built and tested.

### Guidelines for Economical Design

The following guidelines should be considered during the final-design process. These rules are a mix of practical considerations, sound design practices, and common sense. Application of these rules makes the workholder less costly, and improves its efficiency and operation.

*Use Standard Tooling Components.* The economies of standardized parts apply to tooling components as well as to manufactured products. Standard tooling components, readily available from your industrial supplier, include clamps, locators, supports, studs, nuts, pins, and a host of other elements. Most designers would never think of having the shop make cap screws, bolts, or nuts for a workholder. For the same reason, virtually no standard tooling components should be made in-house. The first rule of economic design is: Never build any component you can buy. Commercially available tooling components are manufactured in large quantities for much greater economy.

Labor is usually the largest cost element in the building of any workholder. Standard tooling components are one way to cut labor costs. Always look for new workholding products to make designs simpler and less expensive. Browse through catalogs and magazines to find new products and application ideas. In most cases, the cost of buying a component is less than 20% of the cost of making it.

*Use Prefinished Materials.* Prefinished and preformed materials should be used where possible to lower costs and simplify construction. These materials include precision-ground flat stock, drill rod, structural sections, cast tooling sections, precast tool bodies, tooling plates, and other standard preformed materials. Including these materials in a design both reduces the design time and lowers the labor cost.

*Eliminate Unneeded Finishing Operations.* Finishing operations should never be performed for cosmetic purposes. Making a tool look better can often double the cost of the fixture. Here are a few suggestions to keep in mind with regard to finishing operations:

- Machine only the areas important to the function and operation of the tool. For example, do not machine the edges of a baseplate. Just remove the burrs.
- Harden only those areas of the tool subject to wear.
- Grind only the areas of the fixture where necessary for the operation of the tool.

*Keep Tolerances As Liberal As Possible.* The most-cost-effective tooling tolerance for a locator is approximately 30 to 50% of the workpiece's tolerance. Tighter tolerances normally add extra cost to the tool with little benefit to the process. Where necessary, tighter tolerances can be used, but a tighter tolerance does not necessarily mean a better tool, only a more expensive tool.

*Simplify Tooling Operation.* Elaborate designs often add little or nothing to the function of the jig or fixture. Complex mechanical clamping systems, for example, are often elaborate and unneeded designs. More often, a power clamp could do the same job at a fraction of the cost. Cosmetic details are another example of little gained for the money spent. These details may make the tool look good, but seldom justify the added cost.

Keep the function and operation of a workholder as simple as possible. The likelihood of breakdowns and other problems increases with complex designs. These problems multiply when moving parts are added to the design. Misalignment, inaccuracy, wear, and malfunctions caused by chips and debris can cause many problems in the best tool designs.

Reducing design complexity also reduces misunderstandings between the designer and the machine operator. Whenever possible, a workholder's function and operation should be obvious to the operator without instructions.

## Manual Drawings

Once sketches and the basic workholder design have been completed, final engineering drawings can be prepared. Copies of the engineering drawings, also called shop prints, are used by the toolroom to build the workholder.

Manual drawing is the process of constructing engineering drawings by hand on a drawing board. The easiest way to reduce drawing time is by simplifying the drawing. Words or symbols should be used in place of drawn details where practical. All extra or unnecessary views, projections, and details should be eliminated from the drawing. This cuts the time spent drawing details that add little to the meaning of the drawing.

Drawing a complete clamp assembly, for example, adds very little to the complete design. Simply showing the nose of the clamp in its proper relation to the workpiece, along with specifying its part number, conveys the same information in a fraction of the time.

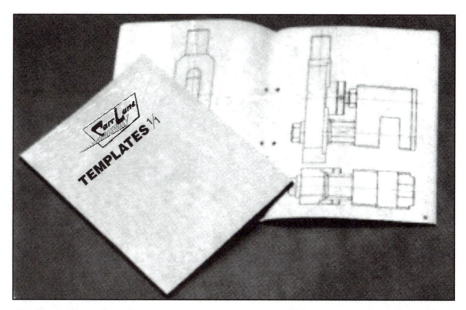

**Figure 2-5**. *Tracing templates can help reduce drawing time when manually drawing standard components.*

For drawings where more detail is required, use tracing templates to reduce drawing time. These templates, shown in Figure 2-5, show most standard components in several views. If necessary, these templates may be enlarged or reduced on a copier to any scale needed for a drawing.

Once the proper template is selected, simply slip it up under the drawing sheet and align it with the drawing. When the template is properly positioned, tape it down and trace the component on the drawing sheet. Tracing templates save drawing time and improve the quality of the drawing.

## Computer-Aided Design

Computers are rapidly replacing drawing boards as the preferred tool for preparing engineering drawings. Almost every area of design is affected by the computer. Computers, from large mainframes to microcomputers, are becoming standard equipment in many design departments.

A standard tooling library, shown in Figure 2-6, is often used to add the fixturing components and elements to the tool drawing. Using

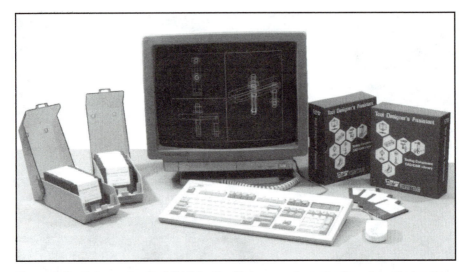

**Figure 2-6.** *Using a standard CAD tooling library can dramatically reduce design time.*

a standard library in designing the workholder dramatically reduces drawing time. All components are drawn to full scale in a variety of views. Each component can be called up from the library and placed on the drawing where it is required.

A CAD system is also sometimes useful during the initial phase of developing numerous tooling options. Computer-aided design is sometimes faster than sketching by hand, especially when detailed cost estimates are required.

## Building and Testing the Workholder

Once drawings have been thoroughly checked, the next step is building the actual workholder. During the building stage, the designer should ensure the toolroom knows exactly what must be done when making the tool. By periodically checking with the toolmakers, the designer can help eliminate any possible misunderstandings and speed the building process. If there are any difficulties with the design, the designer and toolmaker, working together, can solve the problems with a minimum of lost time.

After the tool is completed and inspected, the last step is tool tryout. The workholder is set up on the machine tool and several parts are run. The designer should be on hand to help solve any problems. When the tool proves itself in this phase, it is ready for production.

# 3

# LOCATING
# AND CLAMPING
# PRINCIPLES

Locating and clamping are the critical functions of any workholder. As such, the fundamental principles of locating and clamping, as well as the numerous standard components available for these operations, must be thoroughly understood.

## BASIC PRINCIPLES OF LOCATING

To perform properly, workholders must accurately and consistently position the workpiece relative to the cutting tool, part after part. To accomplish this, the locators must ensure that the workpiece is properly referenced and the process is repeatable.

### Referencing and Repeatability

"Referencing" is a dual process of positioning the workpiece relative to the workholder, and the workholder relative to the cutting tool. Referencing the workholder to the cutting tool is performed by the guiding or setting devices. With drill jigs, referencing is accomplished using drill bushings. With fixtures, referencing is accomplished using fixture keys, feeler gages, and/or probes. Referencing the workpiece to the workholder, on the other hand, is done with locators.

If a part is incorrectly placed in a workholder, proper location of the workpiece is not achieved and the part will be machined incorrectly. Likewise, if a cutter is improperly positioned relative to the fixture, the machined detail is also improperly located. So, in the design of a workholder, referencing of both the workpiece and the cutter must be considered and simultaneously maintained.

"Repeatability" is the ability of the workholder to consistently produce parts within tolerance limits, and is directly related to the referencing capability of the tool. The location of the workpiece relative to the tool and of the tool to the cutter must be consistent. If the jig or fixture is to maintain desired repeatability, the workholder must be designed to accommodate the workpiece's locating surfaces.

The ideal locating point on a workpiece is a machined surface. Machined surfaces permit location from a consistent reference point. Cast, forged, sheared, or sawed surfaces can vary greatly from part to part, and will affect the accuracy of the location.

## The Mechanics of Locating

A workpiece free in space can move in an infinite number of directions. For analysis, this motion can be broken down into twelve directional movements, or "degrees of freedom." All twelve degrees of freedom must be restricted to ensure proper referencing of a workpiece.

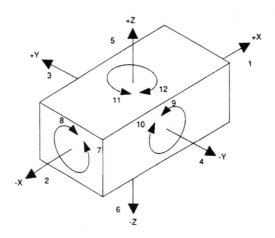

**Figure 3-1.** *The twelve degrees of freedom.*

As shown in Figure 3-1, the twelve degrees of freedom all relate to the central axes of the workpiece. Notice the six axial degrees of freedom and six radial degrees of freedom. The axial degrees of freedom permit straight-line movement in both directions along the three principal axes, shown as x, y, and z. The radial degrees of freedom permit rotational movement, in both clockwise and counterclockwise radial directions, around the same three axes.

The devices that restrict a workpiece's movement are the locators. The locators, therefore, must be strong enough to maintain the position of the workpiece and to resist the cutting forces. This fact also points out a crucial element in workholder design: locators, not clamps, must hold the workpiece against the cutting forces.

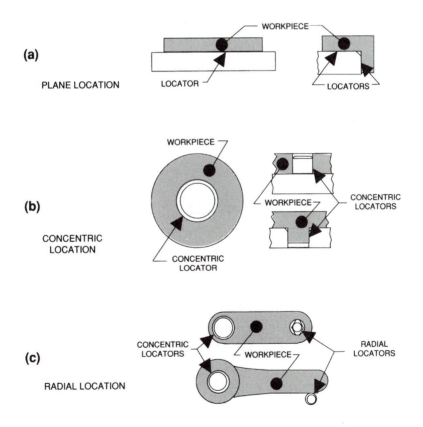

**Figure 3-2.** *The three forms of location: plane, concentric, and radial.*

Locators provide a positive stop for the workpiece. Placed against the stop, the workpiece cannot move. Clamps, on the other hand, rely only upon friction between the clamp and the clamped surface to hold the workpiece. Sufficient force could move the workpiece. Clamps are only intended to hold the workpiece against the locators.

## Forms of Location

There are three general forms of location: plane, concentric, and radial. Plane locators locate a workpiece from any surface. The surface may be flat, curved, straight, or have an irregular contour. In most applications, plane-locating devices locate a part by its external surfaces, Figure 3-2a. Concentric locators, for the most part, locate a workpiece from a central axis. This axis may or may not be in the center of the workpiece. The most-common type of concentric location is a locating pin placed in a hole. Some workpieces, however, might have a cylindrical projection that requires a locating hole in the fixture, as shown in Figure 3-2b. The third type of location is radial. Radial locators restrict the movement of a workpiece around a concentric locator, Figure 3-2c. In many cases, locating is performed by a combination of the three locational methods.

## Locating from External Services

Flat surfaces are common workpiece features used for location. Locating from a flat surface is a form of plane location. Supports are the principal devices used for this location. The three major forms of supports are solid, adjustable, and equalizing, Figure 3-3.

Solid supports are fixed-height locators. They precisely locate a surface in one axis. Though solid supports may be machined directly into a tool body, a more-economical method is using installed supports, such as rest buttons.

Adjustable supports are variable-height locators. Like solid supports, they will also precisely locate a surface in one axis. These supports are used where workpiece variations require adjustable support to suit different heights. These supports are used mainly for cast or forged workpieces that have uneven or irregular mounting surfaces.

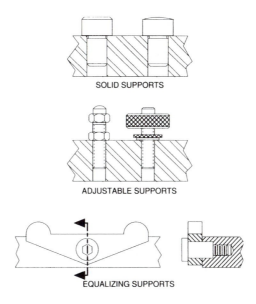

SOLID SUPPORTS

ADJUSTABLE SUPPORTS

EQUALIZING SUPPORTS

**Figure 3-3.** *Solid, adjustable, and equalizing supports locate a workpiece from a flat surface.*

Equalizing supports are a form of adjustable support used when a compensating support is required. Although these supports can be fixed in position, in most cases equalizing supports float to accommodate workpiece variations. As one side of the equalizing support is depressed, the other side raises the same amount to maintain part contact. In most cases adjustable and equalizing supports are used along with solid supports.

Locating a workpiece from its external edges is the most-common locating method. The bottom, or primary, locating surface is positioned on three supports, based on the geometry principle that three points are needed to fully define a plane. Two adjacent edges, usually perpendicular to each other, are then used to complete the location.

The most-common way to locate a workpiece from its external profile is the 3-2-1, or six-point, locational method. With this method, six individual locators reference and restrict the workpiece.

As shown in Figure 3-4, three locators, or supports, are placed under the workpiece. The three locators are usually positioned on the primary locating surface. This restricts axial movement downward, along the -z axis (#6) and radially about the x (#7 and #8) and y (#9 and #10) axes. Together, the three locators restrict five degrees of freedom.

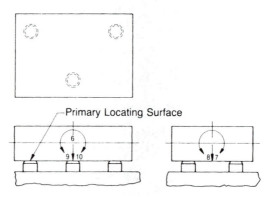

**Figure 3-4.** *Three supports on the primary locating surface restrict five degrees of freedom.*

The next two locators are normally placed on the secondary locating surface, as shown in Figure 3-5. They restrict an additional three degrees of freedom by arresting the axial movement along the +y axis (#3) and the radial movement about the z (#11 and #12) axis.

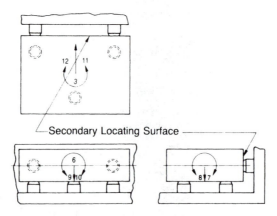

**Figure 3-5.** *Adding two locators on a side restricts eight degrees of freedom.*

The final locator, shown in Figure 3-6, is positioned at the end of the part. It restricts the axial movement in one direction along the -x axis. Together, these six locators restrict a total of nine degrees of freedom. The remaining three degrees of freedom (#1, #4, and #5) will be restricted by the clamps.

Although cylindrical rest buttons are the most-common way of locating a workpiece from its external profile, there are also other devices used for this purpose. These devices include flat-sided locators, vee locators, nest locators and adjustable locators.

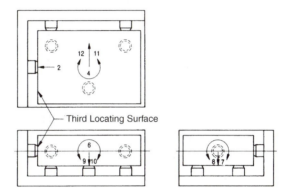

**Figure 3-6.** *Adding a final locator to another side restricts nine degrees of freedom, completing the 3-2-1 location.*

## Locating from Internal Surfaces

Locating a workpiece from an internal diameter is the most-efficient form of location. The primary features used for this form of location are individual holes or hole patterns. Depending on the placement of the locators, either concentric, radial, or both-concentric-and-radial location are accomplished when locating an internal diameter. Plane location is also provided by the plate used to mount the locators.

The two forms of locators used for internal location are locating pins and locating plugs. The only difference between these locators is their size: locating pins are used for smaller holes and locating plugs are used for larger holes.

As shown in Figure 3-7, the plate under the workpiece restricts one degree of freedom. It prevents any axial movement downward,

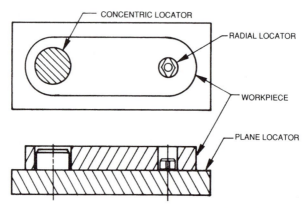

CONCENTRIC LOCATOR

RADIAL LOCATOR

WORKPIECE

PLANE LOCATOR

**Figure 3-7.** *Two locating pins mounted on a plate restrict eleven-out-of-twelve degrees of freedom.*

along the -z (#6) axis. The center pin, acting in conjunction with the plate as a concentric locator, prevents any axial or radial movement along or about the x (#1, #2, #7, and #8) and y (#3, #4, #9, and #10) axes. Together, these two locators restrict nine degrees of freedom. The final locator, the pin in the outer hole, is the radial locator that restricts two degrees of freedom by arresting the radial movement around the z (#11 and #12) axis. Together, the locators restrict eleven degrees of freedom. The last degree of freedom, in the +z direction, will be restricted with a clamp.

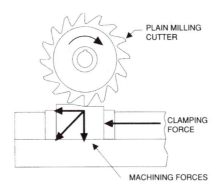

PLAIN MILLING
CUTTER

CLAMPING
FORCE

MACHINING FORCES

**Figure 3-8.** *Cutting forces in a milling operation should be directed into the solid jaw and base of the vise.*

## Analyzing Machining Forces

The most-important factors to consider in fixture layout are the direction and magnitude of machining forces exerted during the operation. In Figure 3-8, the milling forces generated on a workpiece when properly clamped in a vise tend to push the workpiece down and toward the solid jaw. The clamping action of the movable jaw holds the workpiece against the solid jaw and maintains the position of the part during the cut.

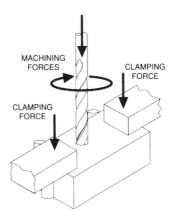

**Figure 3-9.** *The primary cutting forces in a drilling operation are directed both downward and radially about the axis of the drill.*

Another example of cutting forces on a workpiece can be seen in the drilling operation in Figure 3-9. The primary machining forces tend to push the workpiece down onto the workholder supports. An additional machining force acting radially around the drill axis also forces the workpiece into the locators. The clamps that hold this workpiece are intended only to hold the workpiece against the locators and to maintain its position during the machining cycle. The only real force exerted on the clamps occurs when the drill breaks through the opposite side of the workpiece, the climbing action of the part on the drill. The machining forces acting on a correctly designed workholder actually help hold the workpiece.

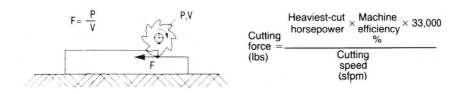

**Figure 3-10.** *A simple formula to estimate the magnitude of cutting forces on the workpiece.*

An important step in most fixture designs is looking at the planned machining operations to estimate cutting forces on the workpiece, both magnitude and direction. The "estimate" can be a rough guess based on experience, or a calculation based on machining data. One simple formula for force magnitude, shown in Figure 3-10, is based on the physical relationship:

$$\text{Force} = \frac{\text{Power}}{\text{Velocity}}$$

Please note: "heaviest-cut horsepower" is not total machine horsepower; rather it is the maximum horsepower actually used during the machining cycle. Typical machine efficiency is roughly 75% (.75). The number 33,000 is a units-conversion factor.

The above formula only calculates force magnitude, not direction. Cutting force can have x-, y-, and/or z-axis components. Force direction (and magnitude) can vary drastically from the beginning, to the middle, to the end of the cut. Figure 3-11 shows a typical calculation. Intuitively, force direction is virtually all horizontal in this example (negligible z-axis component). Direction varies between the x and y axes as the cut progresses.

**Figure 3-11.** *Example of a cutting force calculation.*

## LOCATING GUIDELINES

No single form of location or type of locator will work for every workholder. To properly perform the necessary location, each locator must be carefully planned into the design. The following are a few guidelines to observe in choosing and applying locators.

### Positioning Locators

The primary function of any locator is to reference the workpiece and to ensure repeatability. Unless the locators are properly positioned, however, these functions cannot be accomplished. When positioning locators, both relative to the workholder and to the workpiece, there are a few basic points to keep in mind.

Whenever practical, position the locators so they contact the workpiece on a machined surface. The machined surface not only provides repeatability but usually offers a more-stable form of location. The workpiece itself determines the areas of the machined surface used for location. In some instances, the entire surface may be machined. In others, especially with castings, only selected areas are machined.

The best machined surfaces to use for location, when available, are machined holes. As previously noted, machined holes offer the most-complete location with a minimal number of locators. The next configuration that affords adequate repeatability is two machined surfaces forming a right angle. These characteristics are well suited for the six-point locational method. Regardless of the type or condition of the surfaces used for location, however, the primary requirement in the selection of a locating surface is repeatability.

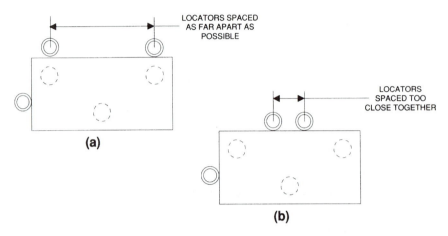

**Figure 3-12.** *Locators should be spaced as far apart as practical to compensate for slight irregularities and for maximum stability.*

To ensure repeatability, the next consideration in the positioning of locators is the spacing of the locators themselves. As a rule, space locators as far apart as practical. This is illustrated in Figure 3-12. Both workpieces shown here are located with the six-point locating method. The only difference lies in the spacing of the locators. In the part shown at (b), both locators on the back side are positioned close to each other. In the part at (a), these same locators are spaced further apart. The part at (a) is properly located; the part at (b) is not. Spacing the locators as far apart as practical compensates for irregularities in either the locators or the workpiece. Its also affords maximum stability.

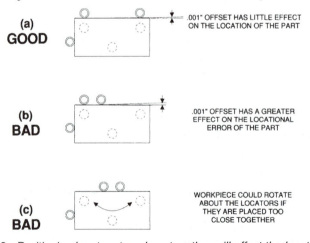

**Figure 3-13.** *Positioning locators too close together will affect the locational accuracy.*

The examples in Figure 3-13 show conditions that may occur when locators are placed too close together if the center positions of the locators are misaligned by .001". With the spacing shown at (a), this condition has little effect on the location. But if the locating and spacing were changed to that shown at (b), the .001" difference would have a substantial effect. Another problem with locators placed too close together is shown at (c). Here, because the locators are too closely spaced, the part can wobble about the locators in the workholder.

## Controlling Chips

The final consideration in the placement of locators involves the problem of chip control. Chips are an inevitable part of any machining operation and must be controlled so they do not interfere with locating the workpiece in the workholder. Several methods help minimize the chip problem. First, position the locators away from areas with a high concentration of chips. If this is not practical, then relieve the locators to reduce the effect of chips on the location. In either case, to minimize the negative effects of chips, use locators that are easy to clean, self-cleaning, or protected from the chips. Figure 3-14 shows several ways that locators can be relieved to reduce chip problems.

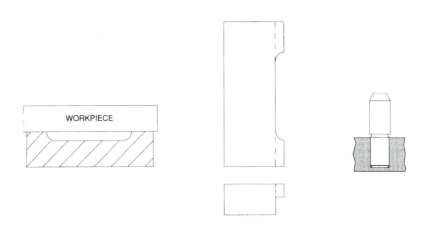

**Figure 3-14.** *Locators should be relieved to reduce locational problems caused by chips.*

Coolant build-up can also cause problems. Solve this problem by drilling holes, or milling slots, in areas of the workholder where the coolant is most likely to build up. With some workholders, coolant-drain areas can also act as a removal point for accumulated chips.

When designing a workholder, always try to minimize the chip problem by removing areas of the tool where chips can build up. Omit areas such as inside corners, unrelieved pins, or similar features from the design. Chip control must be addressed in the design of any jig or fixture.

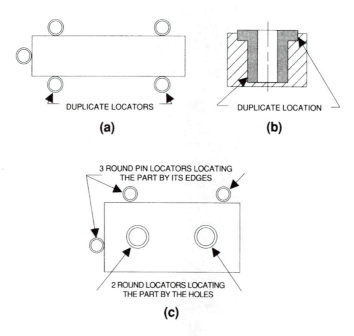

**Figure 3-15.** *Examples of redundant location.*

## Avoiding Redundant Location

Another condition to avoid in workholder design is redundant, or duplicate, location. Redundant locators restrict the same degree of freedom more than once. The workpieces in Figure 3-15 show several examples. The part at (a) shows how a flat surface can be redundantly located. The part should be located on only one, not both, side surfaces. Since the sizes of parts can vary, within their tolerances, the likelihood of all parts resting simultaneously on both surfaces is remote. The example at (6) points out the same problem with concentric diameters. Either diameter can locate the part, but not both.

The example at (c) shows the difficulty with combining hole and surface location. Either locational method, locating from the holes or locating from the edges, works well if used alone. When the methods are used together, however, they cause a duplicate condition. The condition may result in parts that cannot be loaded or unloaded as intended.

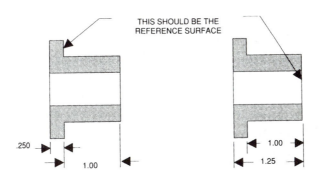

**Figure 3-16.** *The best locating surfaces are often determined by the way that the part is dimensioned.*

Always avoid redundant location. The simplest way to eliminate it is to check the shop print to find which workpiece feature is the reference feature. Often, the way a part is dimensioned indicates which surfaces or features are important. As shown in Figure 3-16, since the part on the left is dimensioned in both directions from the underside of the flange, use this surface to position the part. The part shown to the right, however, is dimensioned from the bottom of the small diameter. This is the surface that should be used to locate the part.

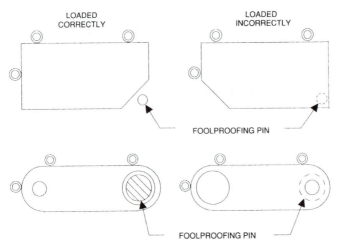

**Figure 3-17.** *Foolproofing the location prevents improper workpiece loading.*

## Preventing Improper Loading

Foolproofing prevents improper loading of a workpiece. The problem is most prevalent with parts that are symmetrical or located concentrically. The simplest way to foolproof a workholder is to position one or two pins in a location that ensures correct orientation, Figure 3-17. With some workpieces, however, more-creative approaches to foolproofing must be taken.

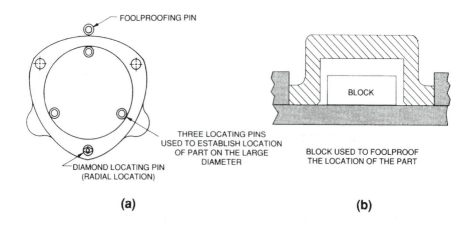

**Figure 3-18.** *Simple pins or blocks are often used to foolproof the location.*

Figure 3-18 shows ways to foolproof part location. In the first example, shown at (a), an otherwise-nonfunctional foolproofing pin ensures proper orientation. This pin would interfere with one of the tabs if the part were loaded any other way. In the next example, shown at (b), a cavity in the workpiece prevents the part from being loaded upside-down. Here, a block that is slightly smaller than the opening of the part cavity is added to the workholder. A properly loaded part fits over the block, but the block keeps an improperly loaded part from entering the workholder.

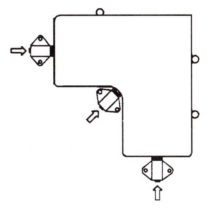

**Figure 3-19.** *Spring-loaded locators help ensure the correct location by pushing the workpiece against the fixed locators.*

## Using Spring-Loaded Locators

One method to help ensure accurate location is the installation of spring-loaded buttons or pins in the workholder, Figure 3-19. These devices are positioned so their spring force pushes the workpiece against the fixed locators until the workpiece is clamped. These spring-loaded accessories not only ensure repeatable locating but also make clamping the workpiece easier.

## Determining Locator Size and Tolerances

The workpiece itself determines the overall size of a locating element. The principle rule to determine the size of the workpiece locator is that the locators must be made to suit the MMC (Maximum-Material Condition) of the area to be located. The MMC of a feature is the size of the feature where is has the maximum amount of material. With external

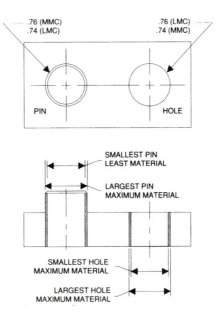

**Figure 3-20.** *Locator sizes are always based on the maximum-material condition of the workpiece features.*

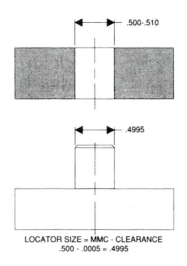

**Figure 3-21.** *Determining the size of a single locating pin based on maximum-material conditions.*

features, like shafts, the MMC is the largest size within the limits. With internal features, like holes, it is the smallest size within the limits. Figure 3-20 illustrates the MMC sizes for both external and internal features.

Sizing cylindrical locators is relatively simple. The main considerations are the size of the area to be located and the required clearance between the locator and the workpiece. As shown in Figure 3-21, the only consideration is to make the locating pin slightly smaller than the hole. In this example, the hole is specified as .500-.510" in diameter. Following the rule of MMC, the locator must fit the hole at its MMC of .500". Allowing for a .0005 clearance between the pin and the hole, desired pin diameter is calculated at .4995". Standard locating pins are readily available for several different hole tolerances, or ground to a specific dimension. A standard 1/2" Round Pin with .4995"-.4992" head diameter would be a good choice.

The general accuracy of the workholder must be greater than the accuracy of the workpiece. Two basic types of tolerance values are applied to a locator: the first are the tolerances that control the size of the locator; the second are tolerances that control its location. Many methods can be used to determine the appropriate tolerance values assigned to a workholder. In some situations the tolerance designation is an arbitrary value predetermined by the engineering department and assigned to a workholder without regard to the specific workpiece. Other tolerances are assigned a specific value based on the size of the element to be located. Although more appropriate than the single-value tolerances, they do not allow for requirements of the workpiece. Another common method is using a set percentage of the workpiece tolerance.

The closer the tolerance value, the higher the overall cost to produce the workpiece. Generally, when a tolerance is tightened, the cost of the tolerance increases exponentially to its benefit. A tolerance twice as tight might actually cost five times as much to produce.

The manufacturability of a tolerance, the ability of the available manufacturing methods to achieve a tolerance, is also a critical factor. A simple hole, for example, if toleranced to ±.050", can be punched. If, however, the tolerance is ±.010", the hole requires drilling. Likewise, if the tolerance is tightened to ±.002", the hole then requires drilling and reaming. Finally, with a tolerance of ±.0003", the hole must be drilled, reamed, and lapped to ensure the required size.

One other factor to consider in the manufacturability of a tolerance is whether the tolerance specified can be manufactured within the capability of the toolroom. A tolerance of .00001" is very easy to indicate on a drawing, but is impossible to achieve in the vast majority of toolrooms.

No single tolerance is appropriate for every part feature. Even though one feature may require a tolerance of location to within .0005", it is doubtful that every tolerance of the workholder must be held to the same tolerance value. The length of a baseplate, for example, can usually be made to a substantially different tolerance than the location of the specific features.

The application of percentage-type tolerances, unlike arbitrary tolerances, can accurately reflect the relationship between the workpiece tolerances and the workholder tolerances. Specification of workholder tolerances as a percentage of the workpiece tolerances results in a consistent and constant relationship between the workholder and the workpiece. When a straight percentage value of 25 percent is applied to a .050" workpiece tolerance, the workholder tolerance is .0125". The same percentage applied to a .001" tolerance is .00025". Here a proportional relationship of the tolerances is maintained regardless of the relative sizes of the workpiece tolerances. As a rule, the range of percentage tolerances should be from 20 to 50 percent of the workpiece tolerance, usually determined by engineering-department standards.

## CLAMPING GUIDELINES

Locating the workpiece is the first basic function of a jig or fixture. Once located, the workpiece must also be held to prevent movement during the operational cycle. The process of holding the position of the workpiece in the jig or fixture is called clamping. The primary devices used for holding a workpiece are clamps. To perform properly, both the clamping devices and their location on the workholder must be carefully selected.

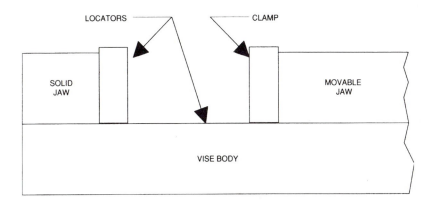

**Figure 3-22.** *A vise contains both locating and clamping elements.*

## Factors in Selecting Clamps

Clamps serve two primary functions. First, they must hold the workpiece against its locators. Second, the clamps must prevent movement of the workpiece. The locators, not the clamps, should resist the primary cutting forces generated by the operation.

*Holding the Workpiece Against Locators.* Clamps are not intended to resist the primary cutting forces. The only purpose of clamps is to maintain the position of the workpiece against the locators and resist the secondary cutting forces. The secondary cutting forces are those generated as the cutter leaves the workpiece. In drilling, for example, the primary cutting forces are usually directed down and radially about the axis of the drill. The secondary forces are the forces that tend to lift the part as the drill breaks through the opposite side of the part. So, the clamps selected for an application need only be strong enough to hold the workpiece against the locators and resist the secondary cutting forces.

The relationship between the locators and clamps can be illustrated with a milling-machine vise. In Figure 3-22, the vise contains both locating and clamping elements. The solid jaw and vise body are the locators. The movable jaw is the clamp. The vise is normally positioned so that the locators resist the cutting forces. Directing the cutting forces into the solid jaw and vise body ensures the accuracy of the machining operation and prevents workpiece movement. In all workholders, it is important to direct the cutting forces into the locators. The movable vise jaw, like other clamps, simply holds the position of the workpiece against the locators.

*Holding Securely Under Vibration, Loading, and Stress.* The next factors in selecting a clamp are the vibration and stress expected in the operation. Cam clamps, for example, although good for some operations, are not the best choice when excessive vibration can loosen them. It is also a good idea to add a safety margin to the estimated forces acting on a clamp.

*Preventing Damage to the Workpiece.* The clamp chosen must also be one that does not damage the workpiece. Damage occurs in many ways. The main concerns are part distortion and marring. Too much clamping force can warp or bend the workpiece. Surface damage is often caused by clamps with hardened or non-rotating contact surfaces. Use clamps with rotating contact pads or with softer contact material to reduce this problem. The best clamp for an application is one that can adequately hold the workpiece without surface damage.

*Improving Load/Unload Speed.* The speed of the clamps is also important to the workholder's efficiency. A clamp with a slow clamping action, such as a screw clamp, sometimes eliminates any profit potential of the workholder. The speed of clamping and unclamping is usually the most-important factor in keeping loading/unloading time to a minimum.

## Positioning the Clamps

The position of clamps on the workholder is just as important to the overall operation of the tool as the position of the locators. The selected clamps must hold the part against the locators without deforming the workpiece. Once again, since the purpose of locators is to resist all primary cutting forces generated in the operation, the clamps need only be large enough to hold the workpiece against the locators and to resist any secondary forces generated in the operation. To meet both these conditions, position the clamps at the most-rigid points of the workpiece. With most workholders, this means positioning the clamps directly over the supporting elements in the baseplate of the work-holder, Figure-3-23a.

In some cases the workpiece must be clamped against horizontal locators rather than the supports, Figure 3-23b. In either case, the clamping force must be absorbed by the locating elements.

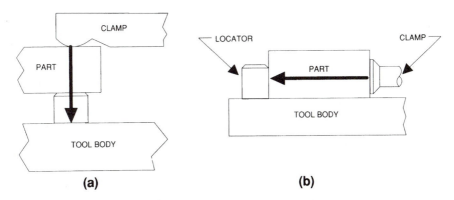

**Figure 3-23.** *Clamps should always be positioned so the clamping force is directed into the supports or locators.*

For workholders with two supports under the clamping area of the workpiece, two clamps should be used — one over each support, Figure 3-24a. Placing only one clamp between the supports can easily bend or distort the workpiece during the clamping operation. When the workpiece has flanges or other extensions used for clamping, an auxiliary support should be positioned under the extended area before a clamp is applied, Figure 3-24b.

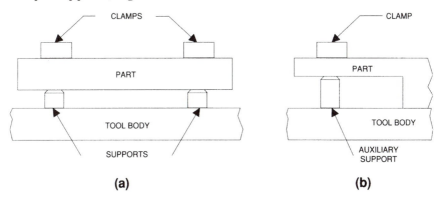

**Figure 3-24.** *The number and position of clamps is determined by the workpiece and its supports.*

Another consideration in positioning clamps is the operation of the machine tool throughout the machining cycle. The clamps must be positioned so they do not interfere with the operation of the machine tool, during either the cutting or return cycle. Such positioning is especially critical with numerically controlled machines. In addition to the cutters, check interference between the clamps and other machine elements, such as arbors, chucks, quills, lathe carriages, and columns.

When fixturing an automated machine, check the complete tool path before using the workholder. Check both the machining cycle and return cycle of the machine for interference between the cutters and the clamps. Occasionally programmers forget to consider the tool path on the return cycle. One way to reduce the chance of a collision and eliminate the need to program the return path is simply to raise the cutter above the highest area of the workpiece or workholder at the end of the machining cycle before returning to the home position.

Most clamps are positioned on or near the top surface of the workpiece. The overall height of the clamp, with respect to the workpiece, must be kept to a minimum. This can be done with gooseneck-type clamps, Figure 3-25. As shown, the gooseneck clamp has a lower profile and should be used where reduced clamp height is needed.

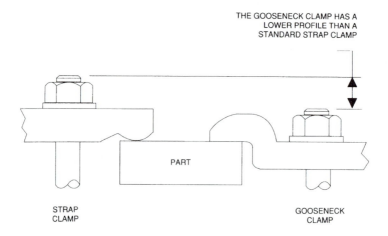

Figure 3-25. *Using gooseneck clamps is one way to reduce the height of the clamps.*

The size of the clamp-contact area is another factor in positioning a clamp. To reduce interference between the clamp and the cutter, keep the contact area as small as safely possible. A small clamping area reduces the chance for interference and also increases the clamping pressure on the workpiece. The overall size of the clamp is another factor to keep in mind. The clamp must be large enough to properly and safely hold the workpiece, but small enough to stay out of the way.

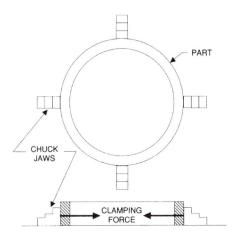

**Figure 3-26.** *Directing the clamping forces against an unsupported area will cause this cylindrical part to deform.*

Once again, the primary purpose of a clamp is to hold the workpiece against the locators. To do this properly, the clamping force should be directed into the locators, or the most-solid part of the workholder. Positioning the clamping devices in any other manner can easily distort or deform the workpiece.

The workpiece shown in Figure 3-26 illustrates this point. The part is a thin-wall ring that must be fixtured so that the internal diameter can be bored. The most-convenient way to clamp the workpiece is on its outside diameter; however, to generate enough clamping pressure to hold the part, the clamp is likely to deform the ring. The reason lies in the direction and magnitude of the clamping force: rather than acting against a locator, the clamping forces act against the spring force of the ring resisting the clamping action. This type of clamping should only be used if the part is a solid disk or has a small-diameter hole and a heavy wall thickness.

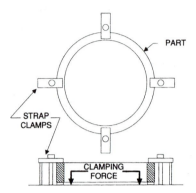

**Figure 3-27.** *Strap clamps eliminate deformation by directing the clamping forces into the supports under the part.*

To clamp this type of part, other techniques should be used. The clamping arrangement in Figure 3-27 shows the workpiece clamped with four strap clamps. The clamping force is directed into the baseplate and not against the spring force of the workpiece. Clamping the workpiece this way eliminates the distortion of the ring caused by the first method.

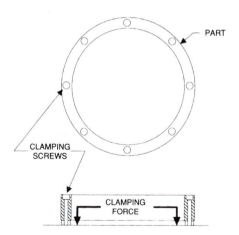

**Figure 3-28.** *When possible, part features such as holes can be used to clamp the part.*

A similar clamping method is shown at Figure 3-28. Here the workpiece has a series of holes around the ring that can be used to clamp the workpiece. Clamping the workpiece in this manner also directs the clamping force against the baseplate of the workholder. This

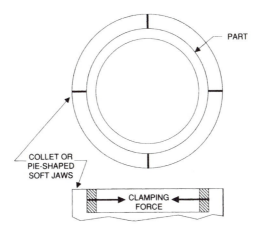

**Figure 3-29.** *When the part can only be clamped on its outside surface, pie-shaped chuck jaws can be used to hold the part and reduce deformation.*

type of arrangement requires supports with holes that permit the clamping screws to clamp through the supports.

If the part can be clamped only on its outside surface, one other method can be used to hold the part: a collet that completely encloses the part. As shown in Figure 3-29, the shape of the clamping contact helps control distortion. Depending on the size of the part, either a collet or pie-shaped soft jaws can be used for this arrangement.

| Stud Size | Recommended Torque* (ft.-lbs.) | Clamping Force (lbs.) | Tensile Force In Stud (lbs.) |
|---|---|---|---|
| #10-32 | 2 | 300 | 600 |
| 1/4-20 | 4 | 500 | 1000 |
| 5/16-18 | 9 | 900 | 1800 |
| 3/8-16 | 16 | 1300 | 2600 |
| 1/2-13 | 38 | 2300 | 4600 |
| 5/8-11 | 77 | 3700 | 7400 |
| 3/4-10 | 138 | 5500 | 11000 |
| 7/8-9 | 222 | 7600 | 15200 |
| 1-8 | 333 | 10000 | 20000 |

*Clean, dry clamping stud torqued to approximately 33% of its 100,000 psi yield strength (2:1 lever ratio).

**Figure 3-30.** *Approximate clamping forces of different-size manual clamp straps with a 2-to-1 clamping-force ratio.*

## Selecting Clamp Size and Force

Calculations to find the necessary clamping force can be quite complicated. In many situations, however, an approximate determination of these values is sufficient. The table in Figure 3-30 shows the available clamping forces for a variety of different-size *manual* clamp straps with a 2-to-1 clamping-force ratio.

Alternatively, required clamping force can be calculated based on calculated cutting forces. A simplified example is shown in Figure 3-31. The cutting force is entirely horizontal, and no workpiece locators are used, so frictional forces alone resist the cutting forces.

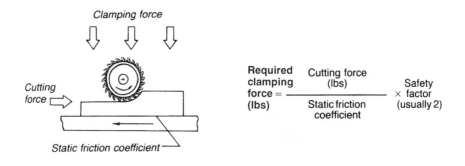

$$\begin{array}{c} \text{Required}\\ \text{clamping}\\ \text{force}\\ \text{(lbs)} \end{array} = \dfrac{\text{Cutting force}}{\text{Static friction}} \times \begin{array}{c}\text{Safety}\\ \text{factor}\\ \text{(usually 2)}\end{array}$$

| Contact surfaces | Friction coefficient (Dry) | Friction coefficient (Lubricated) |
|---|---|---|
| Steel on steel | .15 | .12 |
| Steel on cast iron | .19 | .10 |
| Cast iron on cast iron | .30 | .19 |

**Figure 3-31.** *A simplified clamping-force calculation with the cutting force entirely horizontal, and no workpiece stops (frictional force resists all cutting forces).*

When workpiece locators and multi-directional forces are considered, the calculations become more complicated. To simplify calculations, the worst-case force situation can be estimated intuitively and then treated as a two-dimensional static-mechanics problem (using a free-body diagram). In the example shown in Figure 3-32, the cutting force is known to be 1800 lbs, based on a previous calculation. The workpiece weighs 1500 lbs. The unknown forces are:

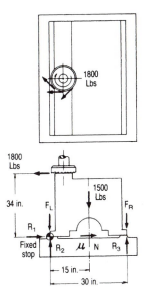

**Figure 3-32.** *A more-complicated clamping-force calculation, using a two-dimensional free-body diagram.*

$F_R$ = Total force from all clamps on right side
$F_L$ = Total force from all clamps on left side
$R_1$ = Horizontal reaction force from the fixed stop
$R_2$ = Vertical reaction force from the fixed stop
$R_3$ = Vertical reaction force on the right side
$N$ = Normal-direction force = $F_L + F_R + 1500$
$\mu$ = Coefficient of friction = .19

The following equations solve the unknown forces assuming that for a static condition:

    1. The sum of forces in the x direction must equal zero.
    2. The sum of forces in the y direction must equal zero.
    3. The sum of moments about any point must equal zero.

At first glance, this example looks "statically indeterminate," i.e., there are five variables and only three equations. But for the minimum required clamping force, $R_3$ is zero (workpiece barely touching) and $F_L$ is zero (there is no tendency to lift on the left side). Now with only three variables, the problem can be solved:

$$\Sigma F_x = 0$$
$$= -1800 + R_1 + (.19)(1500 + F_R)$$
$$\Sigma F_y = 0$$
$$= R_2 - 1500 - F_R$$
$$\Sigma M_\bullet = 0$$
$$= (34)(1800) - (15)(1500) - (30)(F_R)$$

Solving for the variables,

$$F_R = 1290 \text{ lbs}$$
$$R_1 = 1270 \text{ lbs}$$
$$R_2 = 2790 \text{ lbs}$$

In other words, the combined force from all clamps on the right side must be greater than 1290 lbs. With a recommended safety factor of 2-to-1, this value becomes 2580 lbs. Even though $F_L$ (combined force from all the clamps on the left side) equals zero, a small clamping force may be desirable to prevent vibration.

Another general area of concern is maintaining consistent clamping force. Manual clamping devices can vary in the force they apply to parts during a production run. Many factors account for the variation, including clamp position on the workpiece, but operator fatigue is the most-common fault. The simplest and often-best way to control clamping force is to replace manual clamps with power clamps.

The force generated by power clamps is not only constant but also adjustable to suit workpiece conditions. Another benefit of power clamps is their speed of operation: not only are individual power clamps faster than manual clamps, every clamp is activated at the same time.

# 4

# LOCATING DEVICES

Locating is one of the most-important jobs of a jig or fixture. The proper selection of locators contributes to the overall operation and accuracy of the workholder. As discussed in Chapter 3, the locators must first properly position the workpiece, then maintain its location against primary cutting forces throughout the machining cycle.

Most workholders can be built using standard commercial locators. Standard locators both reduce design time and lower the fabrication cost of the workholder. Even if the locator needs to be modified to suit specific part requirements, modifying a commercial locator is usually always less expensive than designing and making the locator in-house.

The terms "locator" and "support" both describe locating devices. Although the term "locator" often describes both locating elements, there are differences between the two, as seen in Figure 4-1. Here the term "support" refers to locators that bear the weight of the workpiece. Supports are generally placed under a part. The term "locator" refers to the elements that position the part on the axes not bearing the weight of the part.

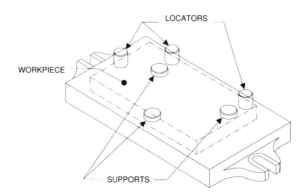

**Figure 4-1.** *The same components can be referred to as "locators" or "supports" depending on the application.*

Locators are also classified by their basic form. Both locators and supports can be grouped into one of two basic categories, fixed and adjustable. Fixed locators are installed in one position on a workholder and are not movable or adjustable. Adjustable locators, while firmly attached to the workholder, are capable of some adjustment to accommodate workpiece variations.

## LOCATING PINS

The most-common form of locator is the pin type, Figure 4-2. Pin locators are available in two basic styles, either plain or with a shoulder.

**Figure 4-2.** *Two common types of locating pin: plain and shoulder-type.*

These locating pins come in a wide range of diameters. They are normally installed by press fitting the pin directly into the tool body.

## Round Locating Pins

Round pins can be used for both internal and external workpiece location. For internal location, the diameter of the pin must match the size of the locating hole. These locators come in many standard sizes, and are readily available ground to special diameters. For external location, the size of the locating pin is not as critical. Here, a standard pin size strong enough to resist machining forces is the best choice.

Plain locating pins are pressed directly into the tool body. They are normally used for workholders in short-to-medium production runs where there is no need for pin replacement. Plain locating pins provide the necessary horizontal location, in the x and y axes, for the workpiece. The vertical location and support, in the z axis, is provided by other supports.

The shoulder-type pins likewise locate the workpiece in the horizontal, x and y, axes. These pins have a shank larger than the head. The purpose of the shoulder is to prevent the pin from being pushed into the tooling plate. Unlike the plain pins, shoulder-type pins are made in two styles: press-fit type and lockscrew type, Figure 4-3. The

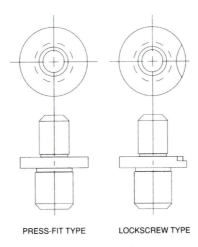

PRESS-FIT TYPE          LOCKSCREW TYPE

**Figure 4-3.** *Shoulder-type locating pins are made in two styles: the press-fit type for permanent installation, and the lockscrew type with a slip fit for renewable installation.*

press-fit type is pressed into the tool body in the same way as a plain-type locating pin. The lockscrew type, however, should be installed with a locating-pin liner bushing.

This liner is pressed into the tool body and affords the locating pin a hardened, wear-resistant mounting hole. The machined recess on the shoulder is for a lockscrew that holds the locating pin in position, Figure 4-4. The liner is primarily intended for workholders in long production runs or for applications where heavy wear is a concern. Locating-pin liners permit the easy and accurate replacement of the locating pins as they wear, without damaging the mounting holes.

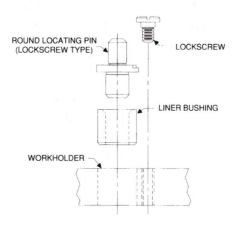

**Figure 4-4.** *Liner bushings and lockscrews are usually used to mount renewable locating pins.*

## Bullet-Nose and Conical Locators

In addition to the round locating pins previously mentioned, other variations are available, including bullet-nose dowels, bullet-nose pins, and cone locator pins, Figure 4-5. These end shapes are mainly for internal location and allow easier loading of workpieces over the pins. Each of the locators is installed by press fitting into the tool body.

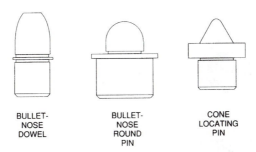

BULLET-
NOSE
DOWEL

BULLET-
NOSE
ROUND
PIN

CONE
LOCATING
PIN

**Figure 4-5.** *These round locating pins have unique head shapes for specialized applications.*

The most-common application for these locating pins is the alignment of workholder elements, rather than locating workpieces. A sandwich jig, for example, is made with two individual plates. Two locating pins ensure the alignment of the top plate to the bottom plate when the jig is assembled. In these cases, the locating pins are aligned with locating bushings, Figure 4-6. These hardened bushings help maintain locational accuracy throughout the life of the workholder.

Bullet-nose round pins are ideal for aligning two pieces of a workholder. The pin's shank diameter and the locating bushing's outside diameter are the same size to allow boring the installation holes in both pieces at the same time, for greater accuracy.

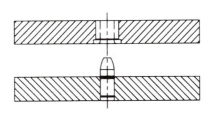

**Figure 4-6.** *With a sandwich jig, bullet-nose dowels and locating bushings ensure the alignment of the top and bottom jig plates.*

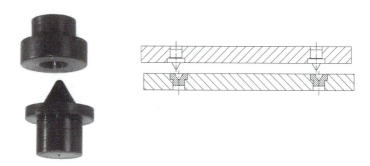

**Figure 4-7.** *Cone locator pins compensate for misalignment to allow quick assembly of two workholder pieces.*

Cone locator pins are used with mating bushings, as shown in Figure 4-7. These medium-accuracy locating pins compensate for a significant amount of misalignment to allow quick assembly of two workholder pieces.

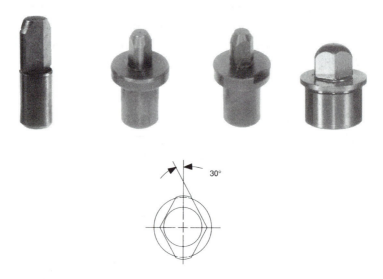

**Figure 4-8.** *Diamond locating pins are relieved to locate in only one axis. They are available in configurations to match round locating pins.*

## Diamond Locating Pins

Another style of locating pin frequently seen in jig-and-fixture design is the diamond, or relieved, locating pin. Like round locating pins, diamond pins are available in either the plain or shoulder-type, as shown in Figure 4-8. These locating pins are the most-common form of relieved locating pin in workholders. To limit the pin's contact area, the diamond locating pin is made with four machined flats. The exact width of the contact area varies with the size of the pin, and is usually equal to one-third of the diameter on each side.

Diamond pins are generally used as shown in Figure 4-9. Here the diamond pin acts as a radial locator to restrict movement of the work-piece around the concentric locator, shown by the round-pin locator. The diamond pin is positioned to restrict the radial movement of the part. Since a diamond pin locates in only one axis, the contact areas of the pin must be positioned as shown. Positioning the pin any other way would allow the part to move about the concentric locator.

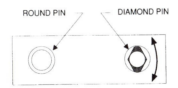

ROUND PIN      DIAMOND PIN

**Figure 4-9.** *Diamond pins are relieved to act only as radial locators, avoiding the redundant location that causes binding during loading.*

## Floating Locating Pins

Another locating pin that corrects slight differences between locating holes is the floating locating pin, Figure 4-10. This pin provides precise location in one axis, but moves up to 1/8" in the perpendicular axis. The body of the locator is referenced to the fixed and movable axes with a roll pin.

The floating locator performs the same function as a diamond pin. Due to the pin's floating movement, however, this locator can be used for parts with looser locational tolerances between the holes. As shown in Figure 4-11, the floating locating pin is often used with a round locating pin.

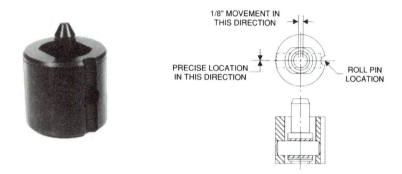

**Figure 4-10.** *A floating locating pin provides precise location in one axis and allows up to 1/8" movement in the perpendicular axis.*

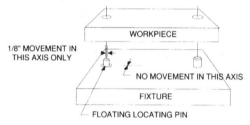

**Figure 4-11.** *Floating locating pins are used with round locating pins to compensate for significant variations in hole spacing.*

## Locating Plugs

Locating plugs, Figure 4-12, are simply large locating pins. Standard locating pins are usually available only up to 1.00" in diameter. Larger plugs are usually pressed into the tool body, then held in

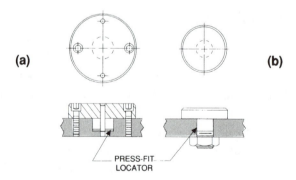

**Figure 4-12.** *Locating plugs can be custom-made to locate workpieces on larger internal diameters.*

place with screws and sometimes dowel pins, as shown at (a). Smaller plugs can be installed with a mounting diameter that has both a press-fit area and threads, as shown at (b). In either case, the press-fit diameter locates the plug in the workholder.

As another option, a series of locating pins can take the place of a locating plug. As shown in Figure 4-13, the pins are positioned at three points, 120 degrees apart, around the internal diameter of the hole. This arrangement is usually much more economical than making a custom plug.

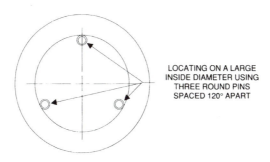

LOCATING ON A LARGE
INSIDE DIAMETER USING
THREE ROUND PINS
SPACED 120° APART

**Figure 4-13.** *Instead of making a special locating plug, three locating pins, spaced 120° apart, can be used for locating large diameters.*

## ADJUSTABLE LOCATORS

Adjustable locators and supports require less precision to mount on the tool body. For a typical workholder, they are mounted at approximately the correct position and then adjusted to the exact location. The adjustable locating buttons shown in Figure 4-14 are a typical example.

The specific design or configuration of an adjustable locator is normally left to the designer. In the design of an adjustable locator, four factors should be considered: 1) the stability and precision of the location; 2) the ability to compensate for variance in workpiece sizes and locating areas; 3) the ability to compensate for wear on the locating element; 4) the ability to maintain the fixed location points.

These design factors, are the same as those required for other types of locational devices, and some are actually easier to achieve with

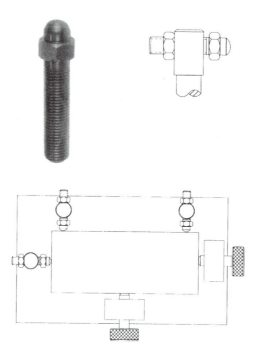

**Figure 4-14.** *Adjustable locating buttons can be used instead of fixed locating pins when some adjustability is needed.*

adjustable locators than with fixed locators. The main disadvantage of adjustable locators is that the operator has the most responsibility of ensuring accuracy and precision, not the tool builder.

To achieve stability and precision, the locator and its mounting device must be large enough to resist all machining and clamping forces. To provide adjustment, the most-popular adjustable locators have a screw thread to control position and accuracy.

The next consideration is compensating for workpiece variations. Size differences are accommodated by setting the locators based on the MMC (Maximum-Material Condition) size of the workpiece. Here adjustable locators have an advantage: part size may vary less within a single production lot, so the locator can be set to a tighter tolerance.

Compensating for locator wear can be controlled with the screw thread or other adjusting device built into the locator. To reduce wear, hardened elements can be used on the contact surfaces.

The final consideration is keeping the adjustable locator in its fixed position during the life of the workholder, or until adjusted for wear. Here the adjustable locator needs a lock to maintain its position. The simplest way is to use a jam nut to hold the position of the locator. Other methods, such as a thread-sealing compound to glue the thread in place, or tack welding, also prevent movement.

## CONICAL LOCATORS

Conical locators, Figure 4-15, as their name implies, have a conical, rather than cylindrical, shape. Like cylindrical locators, small conical locators are usually called pins; larger diameter conical locators are often called plugs. These locators are mainly for locating holes that have large variations in their diameter. Conical locators are often used for cored, cast, or flame-cut holes. The conical locators may be part of a clamp and used on two sides, as shown at (b).

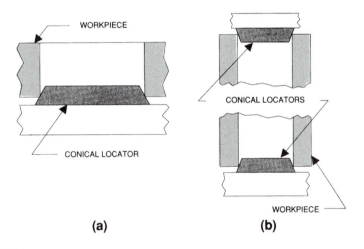

**Figure 4-15.** *Conical locators are often used to locate holes with large diameter variations.*

The main problem with using this type of locator is in the position of the workpiece relative to the workholder. The locating hole can contact the locator at any point on its conical surface. So slight differences in the hole diameter affect the workpiece height, as illustrated in Figure 4-16. This vertical height variation must be considered in the design of the workholder.

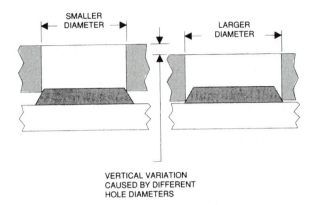

**Figure 4-16.** *The height of a workpiece loaded on a conical locator is affected by the diameter of the locating hole.*

Conical locators can also serve as internal locators, contacting a workpiece on an external surface. As shown in Figure 4-17, a conical locator contacts the external surface of a cylindrical workpiece, its outside diameter.

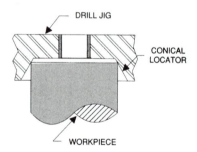

**Figure 4-17.** *Although usually used for locating holes, conical locators can also locate parts on external diameters.*

## VEE LOCATORS

Vee locators are a specialized form of locating element. They are used mainly for round or cylindrical workpieces. The two basic styles of vee locators are the vee pad and the vee block, Figure 4-18.

Vee locators have many applications, as shown in Figure 4-19. The vee accurately locates and centralizes a round workpiece, or a workpiece with radiused ends. Vee pads are well suited for corner mounting a square or rectangular workpiece. The corner selected for the location

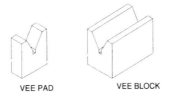

VEE PAD          VEE BLOCK

**Figure 4-18.** *Vee pads and vee blocks are two basic styles of vee locators.*

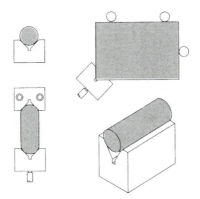

**Figure 4-19.** *Vee locators can be used to locate a variety of part shapes.*

should be a machined right angle to ensure accuracy. Vee blocks, readily available up to 18" long, are used mainly for cylindrical shafts or bars.

The normal angle of vee-type locators is 90°, for stable and consistent location. The angle may be changed, however, when a larger-capacity vee is needed in an area with limited space. A vee with a 120° angle increases the diameter capacity without greatly increasing overall size, Figure 4-20.

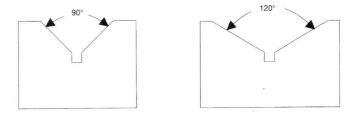

**Figure 4-20.** *Vee locators normally have a 90° angle for the best accuracy and stability, but 120° locators may be used for locating larger diameters where space is limited.*

Vee locators centrally locate the workpiece within the locator. This can result in a locational problem if the vee locator is not properly oriented, as shown in Figure 4-21.

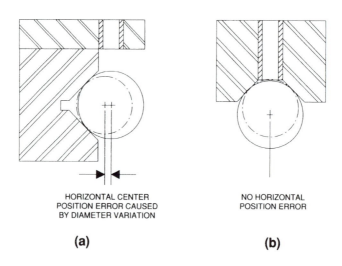

HORIZONTAL CENTER
POSITION ERROR CAUSED
BY DIAMETER VARIATION

**(a)**

NO HORIZONTAL
POSITION ERROR

**(b)**

**Figure 4-21**. *The orientation of a vee block affects its locating accuracy.*

Always position the vee locator to minimize the effect of the workpiece's diameter on the accuracy of the machining operation. As shown at (a), if the vee locator were placed on either side of the workpiece, diameter variations, even within allowable tolerances, could shift the workpiece and cause misalignment of its center. If, however, the vee locator is positioned on the top or bottom of the workpiece, as shown at (b), workpiece-size variation has no effect on the relative position of the workpiece with regard to the cutting tool.

## NESTING LOCATORS

Nesting locators are usually the most-restrictive way to locate a workpiece. These locators normally should be avoided due to their redundant location; however, for complex castings without a machined locating surface, nesting locators are sometimes the only choice.

A nesting locator either partially or completely encases the periphery of a workpiece. It may be machined, cast, or constructed with dowel pins. A machined nest, Figure-4-22, offers complete contact with the part, but is usually very difficult to make.

Cast nests are generally used for complex shapes or for nests that hold parts with irregular locating surfaces. In Figure 4-23, a nest is cast to suit the three-dimensional shape of the part. Casting a nest requires either an epoxy-resin material or a low-melting-point alloy. In either case, a cast nest conforms very well to even the most-intricate part

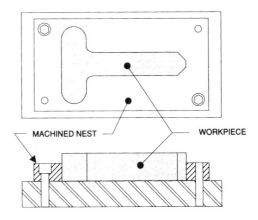

**Figure 4-22.** *A nesting locator completely locates the outside surfaces of a workpiece. It can be machined if the part shape is simple.*

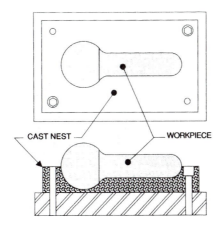

**Figure 4-23.** *For irregular or complex shapes, a nesting locator can also be cast from plastic compounds or low-melting-point alloys.*

shapes. In addition, since the nest is cast, the time and expense involved is only a fraction of that to machine a nest.

It is generally best to make any nest separate from the tool body. The nest is then mounted to the tool body with screws and dowels. When the nest is part of the tool body, problems may result after the workholder begins to wear. If more than one nest is made at a time, the cost of replacing nests is greatly reduced. As the need arises to change the nest, a backup nest is ready to mount on the workholder.

## SUPPORTS

Work supports, like other locators, are made in many forms to suit varied applications. The specific choice of work supports is usually determined by the kind of support required, fixed or adjustable.

### Rest Buttons and Plates

Rest buttons and plates are the primary devices for supporting a workpiece in a jig or fixture. As shown in Figure 4-24, standard rest buttons are available with either flat or spherical contact surfaces in a wide range of sizes. The buttons are typically hardened to reduce wear on the contact surface. Rest buttons with a flat contact are used when a small-area contact on the workpiece is needed. The spherical-style button is best when point contact is preferred.

| REST BUTTON | SPHERICAL-RADIUS LOCATOR BUTTON | SCREW REST BUTTON | JIG REST BUTTON | SCREW REST PAD |

**Figure 4-24.** *Standard forms of rest buttons.*

Rest buttons are usually either press fit, threaded, or screw mounted, as shown in Figure 4-25. The head height of a rest button is closely controlled to ensure accuracy.

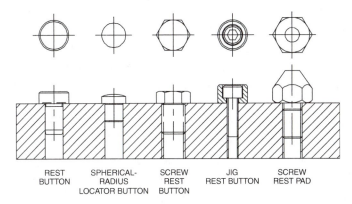

| REST BUTTON | SPHERICAL-RADIUS LOCATOR BUTTON | SCREW REST BUTTON | JIG REST BUTTON | SCREW REST PAD |

**Figure 4-25.** *Rest buttons are usually either press fit, threaded, or screw mounted.*

Rest plates, like rest buttons, are also available in several sizes and thicknesses, Figure 4-26. The plates act in the same way as rest buttons, but are mainly for larger heavier or workpieces. The plates are installed with socket-head cap screws rather than with a shank-type mount. The large pad area reduces contact pressure on the part.

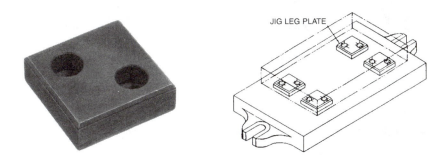

JIG LEG PLATE

**Figure 4-26.** *Jig leg plates support larger or heavier workpieces.*

For most applications, rest buttons and plates are much better than locating surfaces machined directly into the tool body. Since the buttons and plates are mounted to the tool body, they are easily replaced when they wear. Workholders with locators machined directly into the tool body, on the other hand, are not as easily repaired.

The overall accuracy of rest buttons and pads is sufficient for all but the most-critical applications. Typically rest buttons and plates are accurate to -.000/+.001". In cases where more accuracy is required, locating surfaces can be reground. The simplest way to ensure that the heights of the rest buttons or pads are all the same is to install them, then take a light pass across the contact surfaces with a surface grinder (rest buttons are typically through-hardened to allow regrinding). This procedure works only on locators with a flat contact, not with a spherical contact.

Another form of rest pad is the swivel contact bolt, Figure 4-27. These supports have a precision swivel ball that adjusts to curved, sloped, or uneven surfaces to provide full contact. Swivel bolts are also available with serrated round gripper inserts (hardened tool steel or carbide) for additional holding force. Serrated grippers slightly penetrate the workpiece surface to provide significantly more resistance to sliding than from static friction alone.

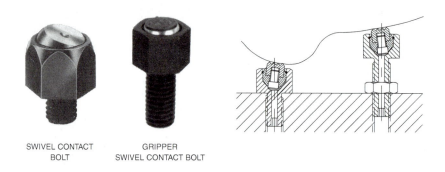

SWIVEL CONTACT
BOLT

GRIPPER
SWIVEL CONTACT BOLT

**Figure 4-27.** *Swivel contact bolts are rest pads that adjust to uneven surfaces. They are also available with round gripper inserts to increase holding force.*

## Jig Feet

Jig feet are devices that locate the entire workholder, a jig, rather than a workpiece. Two common styles are drill-jig feet and double-end feet. Drill-jig feet, Figure 4-28, are mounted on the bottom side of a jig to provide a stable, elevated base for the jig. The feet are mounted on the tool body with a socket-head cap screw, then doweled in place to prevent rotation.

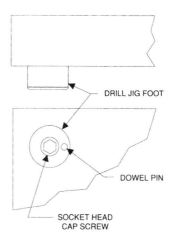

DRILL JIG FOOT

DOWEL PIN

SOCKET HEAD
CAP SCREW

**Figure 4-28**. *Drill jig feet are mounted on the tool body to provide a stable elevated base for the jig.*

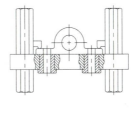

**Figure 4-29**. *Double-end jig feet are available in many different lengths, and are used for jigs that are turned over.*

The double-end jig feet are designed for jigs that are normally used on two sides. The feet are installed on the tool body with studs. As shown in Figure 4-29, one jig foot is installed on one side of the jig and a second foot is mounted on the other side. A stud threaded into both feet securely holds the feet together. Double-end jig feet are generally found in jigs that perform operations on two opposite sides of a part. They are available in several lengths, and different lengths are often used together. Double-end jig feet are also useful for jigs that are flipped over to remove the workpiece. Such construction permits the jig to be flipped over for part loading without resting the jig on the heads of the drill bushings.

For applications where only additional height is needed and the jig will not be flipped over, as with some table jigs, only one set of jig feet is needed, as seen in Figure 4-30. Rest buttons and plates are also frequently used as jig feet.

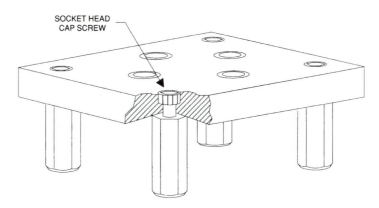

SOCKET HEAD
CAP SCREW

**Figure 4-30.** *A single set of jig feet may also be used for table jigs where only additional height is needed.*

## Screw Jacks

One other device that frequently supports workpieces is the screw jack. Screw jacks are a form of adjustable support. This type of support can be used with almost any workpiece, but it is most-common with cast or forged parts.

Screw jacks, the most-common adjustable support, rely on a screw thread for adjustment. Figure 4-31 shows three variations. The first is the adjustable locating button. The locator has a hexagonal area for adjusting the locator with a wrench. A lock nut prevents movement once the proper height is set.

The other two styles rely on a knurled head for adjustments. The knurled screw jack is a general-purpose device used in applications where the contact force of the support will not move the part. When a specific contact force is desired, however, the torque screw jack should be used. This support is well suited for parts with thin cross sections where excessive pressure can distort the part. Once again, a check nut or lock nut is used to prevent movement once the support height is set.

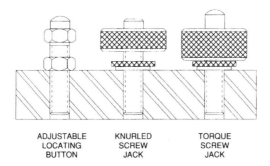

ADJUSTABLE     KNURLED     TORQUE
LOCATING      SCREW      SCREW
BUTTON        JACK        JACK

**Figure 4-31.** *Screw jacks are a common type of adjustable support that use a screw thread for adjustment.*

A fourth style of screw jack is the heavy-duty screw jack shown in Figure 4-32. This screw jack differs from the others in both its construction and the way it is mounted. The unit is mounted on, instead of threaded into, the baseplate. Heavy-duty screw jacks can be used alone or with riser elements to extend the height of the unit. The interchangeable contact tips adapt the jack to various workpiece shapes.

**Figure 4-32.** *The heavy-duty screw jack uses a variety of tips for different part shapes. It can be mounted directly on the baseplate or elevated with riser elements.*

Adjustable supports are normally used as secondary supports in combination with solid supports. In most cases three solid supports are the primary locating points, while adjustable supports provide additional stability, as shown in Figure 4-33.

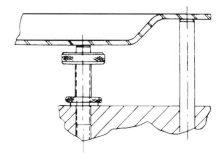

**Figure 4-33**. *Adjustable supports, such as this screw jack, are normally used as supplemental supports along with solid rests.*

## Manual Work Supports

Manual work supports, Figure 4-34, are spring-loaded devices that generally work in combination with solid supports. The support provided is usually secondary rather than primary. Manual work supports provide additional support for thin sections, extended projections, or other workpiece areas that are difficult to support.

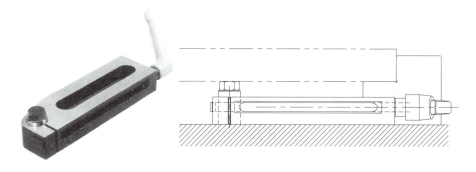

**Figure 4-34**. *Manual work supports have a spring-extended plunger that is locked by tightening the handle. Their low profile allows reaching underneath low workpieces.*

These work supports adjust to different supporting heights with a spring-loaded plunger mechanism. The extended plunger floats while the workpiece is loaded and clamped. Turning the adjustable handle locks the plunger securely. Alternatively, the plunger can also be locked in retracted position for unobstructed loading. A threaded hole in the plunger accepts a locating screw to extend the height of the support.

Manual work supports also dampen vibration. When positioned under large unsupported areas of the part, the supports maintain contact with the part and eliminate much of the chatter caused by vibration.

## Hydraulic Work Supports

Hydraulic work supports, Figure 4-35, also provide adjustable support for irregular or thin areas of a workpiece. Hydraulic supports are more precise and have greater load capacity than manual work supports. The three common forms of hydraulic work supports are the

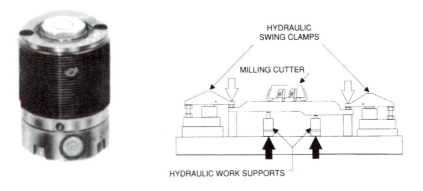

**Figure 4-35.** *Hydraulic work supports automatically adjust to suit the workpiece. Once activated, they move into position and lock automatically to become fixed precision supports.*

spring-extended, fluid-advanced and air-advanced styles, as shown in Figure 4-36. The spring-extended type holds the plunger in contact with the workpiece with an internal spring action. The fluid-advanced type advances the plunger with hydraulic pressure. The air-advanced type moves the plunger into initial contact with the workpiece with air pressure. Once the workpiece is contacted, however, all three styles hydraulically lock the support in position.

Hydraulic work supports do not exert clamping force on the workpiece. Instead, they automatically adjust their plunger height to suit the workpiece. Once positioned, they lock in position and become fixed, precision supports. Several sizes and mounting styles are available to suit almost every application.

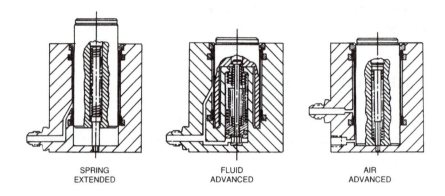

SPRING          FLUID           AIR
EXTENDED        ADVANCED        ADVANCED

**Figure 4-36.** *The three forms of hydraulic work supports are the spring-extended, fluid-advanced and air-advanced styles.*

## Levelers

Levelers are fixturing devices used for a variety of workholding applications. Levelers are often installed in a baseplate to level the complete workholder. They may also level supports for workpieces as well. The major forms of levelers are shown in Figure 4-37. They include leveling feet, jig leveling feet, stud leveling feet, swivel pads, and swivel nuts.

LEVELING FEET        JIG LEVELING FEET        STUD LEVELING
                                             FEET

SWIVEL PADS                                  SWIVEL NUTS

**Figure 4-37.** *Levelers are used both for leveling entire workholders and for supporting workpieces.*

A common feature of the levelers is their swivel contact pad. This is true for all but the swivel nut. The swivel contact pad on each of the levelers has a ball mount, which permits the leveler to be adjusted without rotating the pad. While the leveler is adjusted by turning the threaded body, the pad remains stationary. This prevents workpiece movement and reduces the chance of marring. The swivel nut allows turning a screw body without rotating the contact pad, but does not pivot.

Another style of leveling device is the eccentric leveling lug, Figure 4-38. These levelers most often level tool bodies rather than workpieces. The eccentric leveling lug is first mounted to the tool body through the counterbored mounting hole. The lug is then rotated to the desired height, and the mounting screw is tightened. To permanently fix the location, the lug is doweled in place.

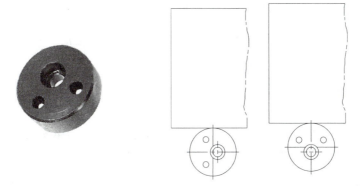

**Figure 4-38.** *Eccentric leveling lugs use a rotary eccentric action for leveling. Once set, they are fixed in place with dowel pins.*

## SPRING-LOADED DEVICES

In addition to fixed and adjustable locators, spring-loaded devices are also used with many jigs and fixtures. Spring-loaded devices include spring locating pins, spring stop buttons, ball plungers, and spring plungers. They play an important role in the operation of many workholders.

### Spring Locating Pins

No matter how well a locating system is designed, unless the parts are properly positioned against the locators every time, mistakes result. One device that reduces locating errors is the spring locating pin, Figure 4-39.

The spring locating pins push the part against the fixed locators. This ensures proper contact during the clamping operation. Although not actually locating devices, the spring locating pins do help reduce errors by correctly positioning the part against the locators. In addition, the pins also eliminate the need for a "third hand" when clamping some parts. Their small size and compact design make them very useful for smaller parts or confined space. A protective rubber seal around the contact pin helps seal out chips and coolant.

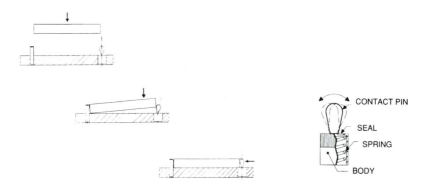

**Figure 4-39.** *Spring locating pins push the workpiece against the fixed locators to ensure proper contact when the workpiece is clamped.*

Spring locator pins can be installed in a hole or mounted in an eccentric liner, Figure 4-40. The liner permits adjustment of the pins to hold parts with looser tolerance, Figure 4-41.

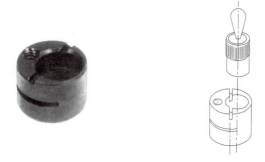

**Figure 4-40.** *Spring locating pins can be mounted directly in a hole or in an eccentric liner.*

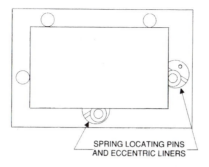

SPRING LOCATING PINS
AND ECCENTRIC LINERS

**Figure 4-41**. *The eccentric liner allows spring-locating-pin adjustment for parts with greater size variations.*

Figure 4-42 shows parts positioned with spring locating pins. These devices can also be used as grippers, or light clamps.

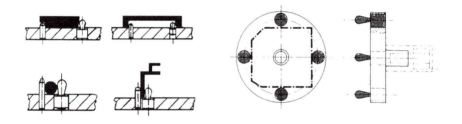

**Figure 4-42.** *Spring locating pins are useful in many workholding applications.*

## Spring Stop Buttons

The spring stop button, Figure 4-43, is another commonly used spring-loaded device. These units work much like the spring pins, but they are designed for larger parts or where more force is needed. Spring stop buttons are made with three different contact faces. The first is a spherical button contact; the other two have a flat contact. The flat-face contacts are made either with or without a tang, Figure 4-44.

SPHERICAL          FLAT          FLAT FACE
BUTTON             FACE          WITH TANG

**Figure 4-43.** *Spring stop buttons are another form of spring-loaded device. They are used to hold larger parts or when more force is needed.*

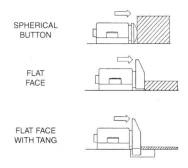

SPHERICAL
BUTTON

FLAT
FACE

FLAT FACE
WITH TANG

**Figure 4-44.** *Spring stop buttons are available with a spherical contact button, flat face, or flat face with tang (for thin sheets).*

## Ball Plungers

Ball plungers are spring-loaded devices. They are used for a variety of workholding applications. As shown in Figure 4-45, the ball

**Figure 4-45.** *A ball plunger and matching ball detent.*

plunger contains a hardened ball as a plunger. In many applications, the ball plunger is combined with a ball detent. The ball detent acts as a hardened locating and referencing device for the ball plunger. The ball plunger has a nylon-type locking element on the threads to prevent backing out after installation.

Ball plungers often locate workholder elements. As shown in Figure 4-46, one application of the units is for locating an indexing arrangement. Here the part is mounted on a pin and held in place with a knurled knob. The first hole is drilled, then the part is rotated until the ball plunger engages the second ball detent. Finally, the mandrel is locked and the second hole is drilled.

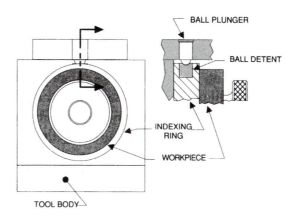

**Figure 4-46.** *Ball plungers and ball detents are often used for indexing operations.*

## Spring Plungers

Spring plungers are similar to ball plungers, except with a longer stroke. Instead of a ball, the spring plunger has a cylindrical plunger with a rounded end. Spring plungers are well suited for a variety of applications, such as the workpiece ejector shown in Figure 4-47.

Spring plungers are available in various lengths and end pressures, in stainless and mild steel. They are also available with Delrin® plungers for applications where marring is a concern. Like the ball plungers, spring plungers have nylon-type locking element on the threads.

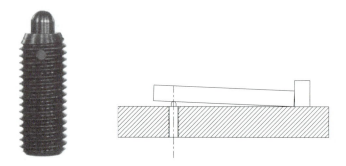

**Figure 4-47.** *Spring plungers have a cylindrical plunger with a rounded end. They are sometimes used as part ejectors.*

## Hand-Retractable Plungers

Hand-retractable plungers, Figure 4-48, are another style of spring plunger. Their purpose is to accurately align workholder elements. Like ball plungers, the hand-retractable plungers often provide indexing, but they engage an indexing hole rather than a ball detent. Such alignment both ensures the correct position and provides a positive lock. As shown in Figure 4-49, these plungers are hand operated and must be pulled back to disengage the indexing hole. When the handle is released, a spring advances the plunger into the hole.

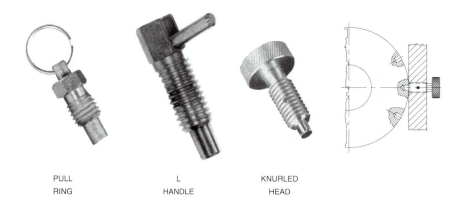

PULL          L          KNURLED
RING     HANDLE     HEAD

**Figure 4-48.** *Hand-retractable plungers are spring-loaded plungers used to accurately align workholder elements in an indexing or referencing hole.*

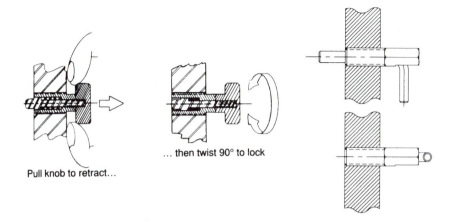

**Figure 4-49.** *Hand-retractable plungers are advanced with a spring action and retracted manually.*

## ALIGNMENT PINS

Alignment pins serve a variety of functions with jigs and fixtures. Most commonly, these devices align the workpiece to the workholder, or position removable workholder elements. Alignment pins come in many types, styles, and sizes for many applications.

### Plain Alignment Pins

Plain alignment pins, Figure 4-50, are the simplest form of alignment pin. The five primary types are L pins, T pins, jig pins, shoulder pins, and clamping pins. Each is made with a precise-diameter locating pin to ensure proper alignment. All but the clamping pin have a bullet-nose end for easy insertion into the mounting holes. The clamping pin has a threaded end to permit a fixed installation. It also comes with a bushing that is positioned and fixed at assembly.

The specific mount for the alignment pins is determined by the application. In some cases the pins are installed in drilled and reamed holes. For more precision or for longer production runs, they are installed in hardened bushings. Slotted locator bushings, Figure 4-51, are used to line up two sets of holes without binding. The design of these bushings is similar to that of floating locating pins and permits

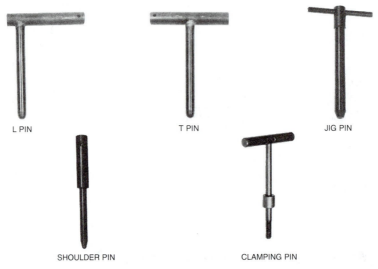

L PIN                    T PIN                    JIG PIN

SHOULDER PIN              CLAMPING PIN

**Figure 4-50.** *Alignment pins are used to accurately line up holes.*

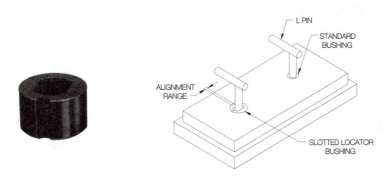

**Figure 4-51.** *Slotted locator bushings are used with alignment pins to line up two sets of holes without binding.*

movement in only one direction. As shown, the bushings are installed and correctly aligned in the mounting plate with a dowel pin.

Slotted bushings are also available with a knurled outside diameter. These bushings can be either cast in place or potted in a plastic compound.

## Locking Alignment Pins

In addition to the plain style, L pins, T pins, and clamping pins are also available with locking bushings. As shown in Figure 4-52, locking-pin bushings are installed in the mounting plate and securely attach the alignment pin to the plate. The spring clip ring in the locking-pin bushing holds the alignment pin in its retracted position by engaging the groove at the end of the pin. The ring also puts spring pressure on the pin at any intermediate position. This is useful for holding the pin in position if the workholder is turned upside down.

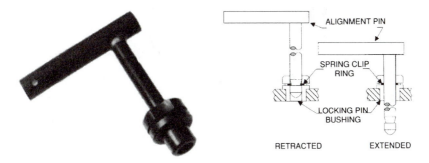

**Figure 4-52**. *Locking alignment pins are positively retained in retracted position by a spring clip.*

## Quick-Release Alignment Pins

Quick-release pins have an integral locking mechanism in the pin itself which retains it in the hole. The simplest quick-release pin is the detent pin, Figure 4-53. It has a simple spring-loaded dual-ball arrangement. The pin is simply pressed in or pulled out, against light spring force which extends the balls.

**Figure 4-53**. *Detent pins are economical locating pins with spring-loaded locking balls.*

Another variation is the ball-lock pin. This group includes single-acting ball-lock pins, double-acting ball-lock pins, adjustable ball-lock pins, and lifting pins. These pins are similar to the detent type, except they have a positive ball-locking mechanism. Single-acting pins, Figure 4-54, are installed and removed by pushing a button to unlock the balls. Double-acting pins, Figure 4-55, are released by either pulling or pushing the handle. Both single- and double-acting ball-lock pins have a

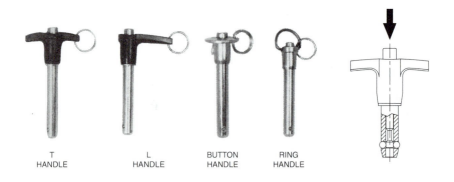

T HANDLE     L HANDLE     BUTTON HANDLE     RING HANDLE

**Figure 4-54.** *Single-acting ball-lock pins are precision alignment pins which are positively locked until released by pressing a button. They are available in four handle styles.*

T HANDLE     RING HANDLE     L HANDLE

**Figure 4-55.** *Double-acting ball-lock pins are released by either pulling or pushing the handle. They are available in three handle styles.*

fixed grip length. When grip length must be fine tuned, the adjustable ball-lock pin, Figure 4-46, can be used. This pin's grip length is adjusted ±1/4" by turning the handle, then locked in place with a knurled lock nut.

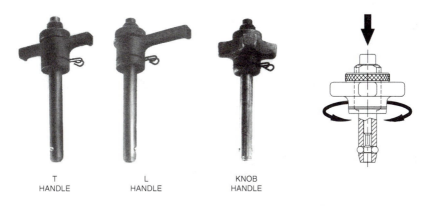

T HANDLE      L HANDLE      KNOB HANDLE

**Figure 4-56**. *Adjustable ball-lock pins are single-acting pins with a variable grip length. They are available in three handle styles.*

Another specialized single-acting ball-lock pin is the lifting pin, Figure 4-57. Lifting pins have a solid, one-piece body with four locking balls and a forged shackle for heavy lifting.

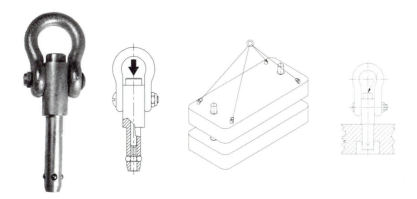

**Figure 4-57**. *Lifting pins are single-acting ball-lock pins designed for heavy lifting, with a one-piece body, four locking balls, and a forged shackle.*

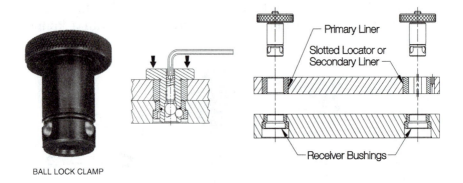

BALL LOCK CLAMP

**Figure 4-58.** *The Ball Lock Mounting System™ incorporates a ball-lock pin that simultaneously locates and clamps, ideal for mounting quick-change tooling on a subplate.*

The Ball Lock Mounting System™, Figure 4-58, is a combination locator and clamp based on ball-lock-pin principles. Each mount consists of three components: (1) a ball-lock pin/clamp with a precisely ground shank; (2) a liner bushing in the top plate, either plain or slotted; (3) a receiver bushing in the subplate. Turning the clamping screw with a hex wrench advances the large center ball, pushing the three clamping balls outward.

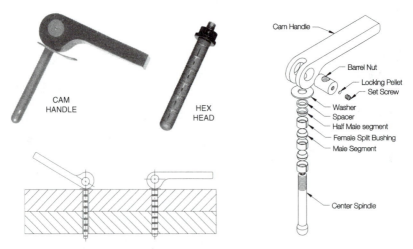

**Figure 4-59.** *Expanding pins are precision, high-strength alignment pins that expand to tighten up hole clearance.*

Yet another type of quick-release pin is the expanding pin, Figure 4-59. Expanding pins are precision alignment pins that expand up to .006" to tighten up hole clearance. The pin is actually an assembly of female split bushings separated by male segments, on a center spindle, that expands when drawn together. This assembly provides excellent shear strength, comparable to a solid pin. Expanding pins are available in two versions, actuated either by cam handle or by tightening with a wrench.

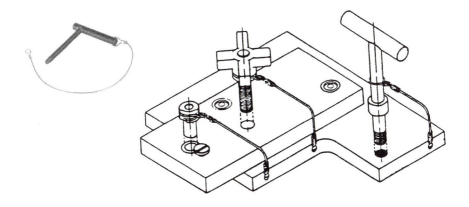

**Figure 4-60.** *Cable assemblies are handy for attaching alignment pins and other removable items to a workholder.*

## Cable Assemblies

Cable assemblies attach alignment pins and other items to a workholder. Attached to the workholder, as shown in Figure 4-60, the items remain handy and less prone to loss or misplacement. Cable assemblies and their individual parts are available in many types, sizes, and lengths to suit almost every application.

# 5

# CLAMPING DEVICES

Securely holding the workpiece is an essential function of any jig fixture. The first step in selecting and applying clamps is understanding their basic actions and the characteristics of efficient clamping. Manually operated clamps can be divided into several basic groups: strap clamps, screw clamps, swing clamps, edge clamps, C clamps, cam clamps, and toggle clamps. Some clamps fit into more than one classification.

## STRAP CLAMPS

A clamp strap is the simplest and most-common clamp. The basic clamp-strap assembly consists of three major elements: a clamp strap, a fastening device, and a heel support, Figure 5-1. Force is applied to the fastening device. The force is then transferred through the strap to the workpiece. The heel support acts as a pivot and support for the back end of the strap.

### Clamp-Strap Operation

All clamp straps work on the mechanical principle of the lever. As shown in Figure 5-2, the three basic styles of strap clamps can be described in terms of lever arrangements, called first-, second-, and

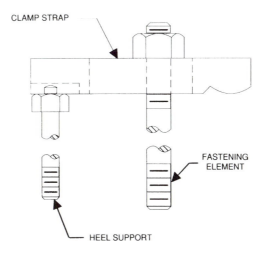

**Figure 5-1.** *The major elements of a clamp strap assembly are the clamp strap, fastening element or assembly, and heel support.*

third-class levers. The classes are not meant to describe importance or preference levels; rather they show distinctions in the mechanical actions of each lever style.

As shown, the major difference between the three strap-clamp variations lies in the arrangement of the three main elements in each lever. These elements are the force-applying device, the workpiece, and the fulcrum.

As shown in Figure 5-2a, a first-class lever has the workpiece at one end, the fulcrum in the center, and the force at the opposite end of the clamp strap. A second-class lever, shown in Figure 5-2b, has the fulcrum at one end, the workpiece in the center, and the force at the other end of the clamp strap. Finally, as shown in Figure 5-2c, a third-class lever places the part at one end of the clamp strap, the force in the center, and the fulcrum at the opposite end. Each of these arrangements is well suited for certain workholding situations. With strap clamps, the third-class-lever arrangement is the most common.

Clamping force on the workpiece depends on the relative position of the workpiece, fastening element, and fulcrum. All strap clamps are

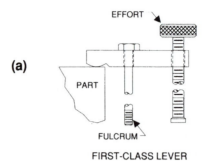

**(a)**

FIRST-CLASS LEVER

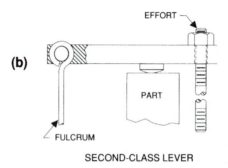

**(b)**

SECOND-CLASS LEVER

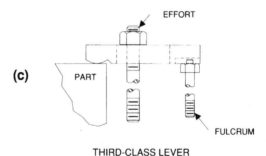

**(c)**

THIRD-CLASS LEVER

**Figure 5-2.** *Strap clamps can be categorized as first-, second-, or third-class levers.*

basically beams that are loaded in bending, as shown in Figure 5-3. The loads on a clamp strap are the applied force $F$, clamping force $P$, and reaction force $R$. The applied force is the force applied by the fastening device. For most calculations, this force is known. The clamping force is the actual force applied to the workpiece during the clamping operation, a fractional portion of the applied force. The reaction force is the force generated on the fulcrum. Changing the positions of the various elements affects the amount of clamping force.

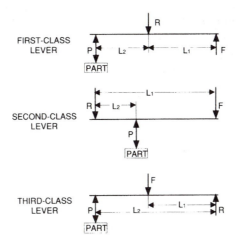

**Figure 5-3.** *Force diagrams show the difference between the three classes of levers.*

The value $L_1$ is the distance between the fulcrum and the applied force. $L_2$ is the distance between the fulcrum and the part-contact point. These values determine the force ratio (mechanical advantage) of a clamping arrangement. The actual clamping force applied to the workpiece with each setup is a proportional amount of the applied force. The ratio of actual clamping force to applied force is equal to the ratio of the two distances:

$$\frac{P}{F} = \frac{L_1}{L_2}$$

With a third-class lever arrangement, as shown in Figure 5-4, the force on the workpiece depends on the position of the stud with

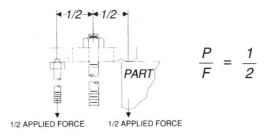

$$\frac{P}{F} = \frac{1}{2}$$

**Figure 5-4.** *Positioning the stud in the center of a third-class lever distributes the force equally between the workpiece and heel support.*

respect to the workpiece and the heel support. If the stud is positioned exactly in the center of the clamp strap, the force generated by the fastener is distributed equally between the workpiece and the heel support. If two workpieces are clamped with a single strap-clamp arrangement, as shown in Figure 5-5, positioning the stud in the center of the clamp strap applies equal holding force to both parts. In Figure 5-6, the

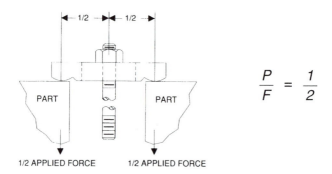

$$\frac{P}{F} = \frac{1}{2}$$

**Figure 5-5.** *When clamping two workpieces, positioning the stud in the center of the clamp strap applies equal holding force to both workpieces.*

fastener is positioned so that only one-third of the clamp-strap length is between the fastener and workpiece, while two-thirds is between the fastener and the heel support. The clamping force on the workpiece with this setup is twice as great as that on the heel support.

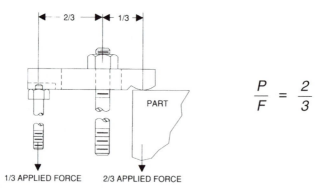

$$\frac{P}{F} = \frac{2}{3}$$

**Figure 5-6.** *Positioning the stud closer to the workpiece generates proportionally more force on the workpiece and less on the heel support.*

## Types of Clamp Straps

Clamp straps come in a variety of forms. Selecting the correct clamp strap for a task is important to a workholder's operation and efficiency. Clamp straps vary both in material and basic form or style. The most-common material for clamp straps is carburized-hardened carbon steel, but forged steel, stainless steel, and aluminum clamp straps are also available.

The specific material used for a particular workholder is normally determined by the application itself. Steel clamp straps are normally used for general-purpose applications. Stainless steel clamp straps are used when more corrosion resistance is needed. Forged steel clamp straps are normally employed for applications where extra toughness is required. Aluminum clamp straps are often used where weight is important or where softer clamps are needed to prevent marring workpiece surfaces.

Clamp straps can also be classified based on the form of the strap. Variations in the end shapes, contact areas, and heel supports are all used to classify clamp straps. Like the material, these factors can greatly influence the selection and application of these clamps.

The most-common clamp-strap shapes are shown in Figure 5-7. They include the plain strap (a), tapered-nose strap (b), wide-nose strap (c), U-shaped strap (d), gooseneck strap (e), and double-end strap (f). The variety of end shapes provides clamp straps tailored to exact clamping requirements and conditions.

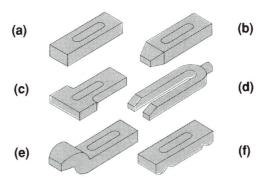

**Figure 5-7.** *Clamp straps are made in a variety of shapes. Most common are the plain, tapered-nose, wide-nose, U-shaped, gooseneck, and double-end straps.*

The workpiece-contact area of the clamp strap is also an important consideration in clamp selection. The contact areas, Figure 5-8, can be either flat or radiused. The radius nose (a) is usually best because it keeps the contact area to a minimum. A small contact area best avoids warping the workpiece. The flat contact (b) offers broader contact with the workpiece, for less contact pressure. A variation of the flat contact is the padded clamp strap (c). Padded straps come with a steel pad, but brass or plastic pads are also available.

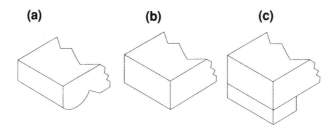

**Figure 5-8.** *Clamp straps are available with different workpiece-contact areas. The most-often used are radiused, flat, and padded.*

In addition to a variety of forms and contact areas, clamp straps are also made with several different types of heel supports. The heel support area of a clamp strap is at the end opposite the contact area. Figure 5-9 shows several heel supports. The most-basic heel support

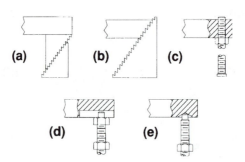

**Figure 5-9.** *Various heel supports are used for clamp straps. The most-common clamp heels are the plain, stepped, tapped, slotted, and drill-spot heels.*

shape is the plain heel (a). The plain heel, as shown here, frequently uses step blocks as a support. A variation of this heel support also takes a step-block approach, the stepped-heel clamp strap (b). This clamp incorporates the same pattern of steps into the end of the strap. Only one step block, rather than two, is required to support the end of the clamp. The tapped-heel clamp strap (c) uses a threaded heel support for adjustment.

The last types of heel support include the slotted-heel type (d) and the drill-spot-heel type (e). The slotted-heel clamp strap has a shallow slot cut into the underside. This slot aligns with a clamp rest screw which adjusts to suit the height of the workpiece. This sliding feature permits the clamps to be easily moved on and off the workpiece, while returning to the same clamping position relative to the workpiece. The drill-spot-heel style has a drill spot on the underside rather than a slot. The drill spot helps position the clamp on the clamp rest.

Although each area of the clamp strap has been discussed separately here, in practice these elements are combined into a variety of standard clamps. The more-common types are as follows:

*Slotted-Heel Clamp Straps.* The slotted-heel clamp strap, Figure 5-10, has a slotted heel to locate the clamp strap on the clamp rest. This slotted heel, in combination with the slotted stud hole, allows the clamp strap to move completely clear of the workpiece for part removal and

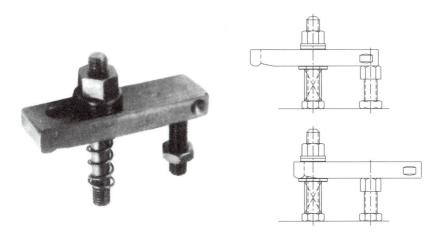

**Figure 5-10.** *The slotted-heel clamp strap uses a slot to guide the strap on the clamp rest. This design permits the clamp strap to be moved completely clear of the workpiece, for easier loading/unloading of workpieces.*

reloading. These clamps are made in many lengths, and in stud sizes from #10-32 through 3/4-10, in either carbon or stainless steel. The contact area of these clamp straps is available in radius-nose, plain-nose, and padded-nose styles.

Variations of this clamp strap are used on high-rise clamps, shown in Figure 5-11. These assemblies can be mounted on riser blocks for tall workpieces. They come in the three basic styles shown here, with a standard slotted-heel clamp strap, a tapped-nose clamp strap, or a gooseneck clamp strap. The tapped-nose version has a tapped hole at the clamping end, for mounting contact elements or adjustable spindles. The pin end, located opposite the tapped end, is for clamping in a horizontal hole or slot. The gooseneck clamp strap is used to lower the height of the fastening element.

**Figure 5-11.** *High-rise clamp assemblies are ideal for clamping taller workpieces. These all-in-one assemblies can be stacked on risers for additional height.*

*Tapped-Heel Clamp Straps.* The tapped-heel clamp strap, Figure 5-12, has a tapped hole in the heel end of the strap. The component installed in this hole can be either an adjustable heel rest or a screw clamp that applies clamping force, Figure 5-13. When used to mount an adjustable heel rest, the complete clamp is a third-class-lever arrangement, but when used to mount a screw clamp, it then employs a first-class-lever action.

Like slotted-heel clamp straps, tapped-heel straps are made in many sizes, of either carbon or stainless steel. The contact area of these clamp straps is also available in radius-nose or padded-nose styles. Two

 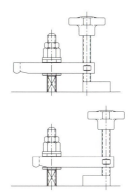

**Figure 5-12.** *The tapped-heel clamp strap has a tapped hole in the heel to mount a clamping screw. A slotted guide block allows the strap to slide straight back and forth.*

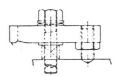 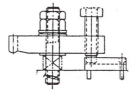

**Figure 5-13.** *Tapped-heel clamp straps can be used with either an adjustable heel rest or a device to apply clamping force.*

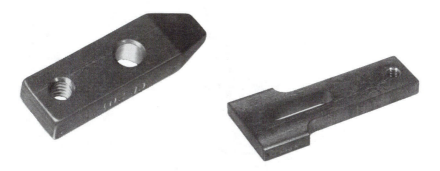

**Figure 5-14.** *Tapped-heel clamp straps are also available with a tapered nose or wide nose.*

**Figure 5-15.** *Tapped-heel clamp straps are also made in either forged steel or aluminum alloy, in straight and gooseneck styles.*

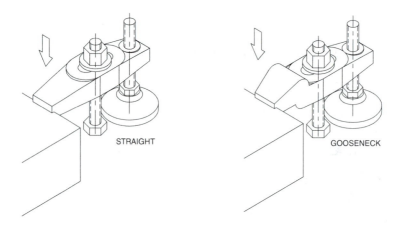

**Figure 5-16.** *Forged tapped-heel clamp straps are available with a locator nose for precise clamp position.*

**Figure 5-17.** *The forged tapped-heel clamp strap often uses a stud leveling foot as a clamp rest.*

other end variations include the tapered-nose and wide-nose styles, Figure 5-14. The tapered nose works well when space is limited. The wide nose is used where a larger contact area is needed either to hold the workpiece or to spread out the clamping force.

*Forged Tapped-Heel Clamp Straps.* The forged tapped-heel clamp strap, Figure 5-15, is a variation of the tapped-heel clamp strap made in either forged steel with a 70,000-psi yield strength, or in aluminum alloy with a 25,000-psi yield strength. Both varieties are made in either straight or gooseneck styles, with either a plain nose or a locator nose. The locator-nose clamps, Figure 5-16, have an accurately machined locating notch to ensure precise, repeatable clamp positioning. This is ideally suited where reduced clamp contact is needed for machining clearance.

Like the tapped-heel clamp strap, these clamp straps also have a tapped hole in the heel end of the strap for a threaded heel support. Though this hole could also be used to mount a screw clamp, more often stud leveling feet are used as clamp rests, Figure 5-17.

*Double-End Clamp Straps.* The double-end clamp strap holds workpieces on both ends. As shown in Figure 5-18, this clamp strap has a radius nose on both ends. This design provides equal clamping force on both ends, even if there are height variations in the workpieces.

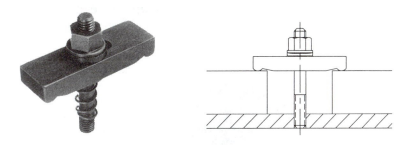

**Figure 5-18.** *The double-end clamp strap is designed to clamp on both ends and applies equal force to both workpieces.*

*Cam-Type Clamp Straps.* Cam-type clamp straps, Figure 5-19, are made with a slotted end to mount a single cam. This design is a first-class lever. The clamp strap is attached to the workholder with a stud assembly and the single cam is mounted at the end. When the cam is depressed, the clamp strap pivots on the stud assembly, which acts as the fulcrum.

**Figure 5-19.** *Cam-type clamp straps use a single cam mounted in the end slot to apply clamping force.*

*Drill-Spot-Heel Clamp Straps.* The drill-spot-heel clamp strap, Figure 5-20, uses a drill spot, rather than a slot, to accurately position the heel support. This design permits the clamp strap to be moved back or swiveled out of the way for loading/unloading.

**Figure 5-20.** *A drill-spot-heel clamp strap positions the heel support in a drill spot rather than a slot.*

*Swivel-Heel Clamp Straps.* The swivel-heel clamp strap uses a variation of the slotted-heel design to position the heel support. The slotted-heel design uses a straight slot, machined parallel to the clamp axis. The swivel-heel design, however, uses a curved slot machined perpendicular to this axis, Figure 5-21. This arrangement allows the clamp strap to be easily swiveled clear of the workpiece for loading/unloading operations.

**Figure 5-21.** *The swivel-heel clamp strap is a curved-slot variation of the slotted-heel design. This allows the strap to swing out of the way for loading.*

*Step Clamps.* Step clamps are one of the more-common clamp strap designs. These clamp straps have a series of serrations, or steps, machined in their heel, Figure 5-22a. These steps are designed to engage a similar set of steps in a matching heel block, Figure 5-22b. This design allows the clamp to be positioned for different workpiece heights.

**Figure 5-22.** *Step-type clamp straps have serrations machined in their heel to engage a matching heel block.*

*Forged U Clamps.* The forged U clamp, Figure 5-23a, is a general-purpose clamp strap. As shown in Figure 5-23b, both the U end and pin end of the forged steel clamp strap can clamp a workpiece. The U end permits the strap to be completely removed when unclamped. The pin end is often used where space is limited or for clamping in horizontal holes.

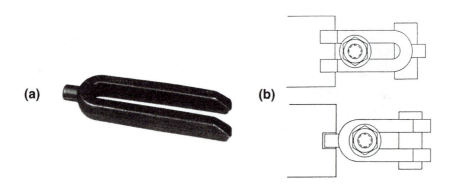

**(a)**    **(b)**

**Figure 5-23.** *Forged U clamp straps can be used for clamping at either the U end or the pin end.*

*Forged Adjustable Clamps.* Forged adjustable clamps, Figure 5-24, are well suited where a heel support could interfere with the setup. These clamps are made as a complete unit and can accommodate a

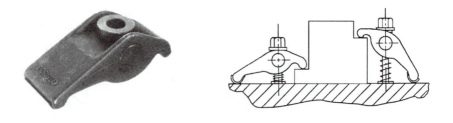

**Figure 5-24.** *Forged adjustable clamps have a built-in heel support.*

variety of workpiece heights. The steel pivot allows the fastening element to securely clamp at all elevations. This clamp is available in either forged steel or aluminum alloy.

*Forged Screw Clamps.* The forged screw clamp, Figure 5-25, is a general-purpose clamping device well suited for a variety of clamping operations. Though designed for a T slot on a machine table, the clamp can also be mounted to a tool body with a single screw. The sliding T handle provides extra leverage for clamping.

**Figure 5-25.** *Forged screw clamps use a clamping screw to apply force.*

## Fastening Elements

The fastening element of a clamp strap is the device that actually applies force to the clamp. The two general types of fastening elements for strap clamps are threaded fasteners and cam handles. Threaded fasteners include a wide variety of bolts, studs, washers, nuts, and knobs.

One note of caution about fastening elements: always make sure that fasteners are specifically made for workholding operations. Standard hardware items are not strong enough for consistently safe clamping. Likewise, many low-cost, cut-rate clamping components will not stand up to the repeated use required in workholding. These items often bend or rupture under the severe conditions imposed on clamps in production applications.

*Studs and Bolts.* Two major types of threaded fasteners used for strap clamps are studs and bolts. Studs are the most-common fastener for strap clamps. One end of the stud is usually mounted in a T nut; the other applies the holding force with a nut, as shown in Figure 5-26a. Alternatives to the stud-and-T-nut combination are T bolts, Figure 5-26b, and T-slot bolts, Figure 5-26c.

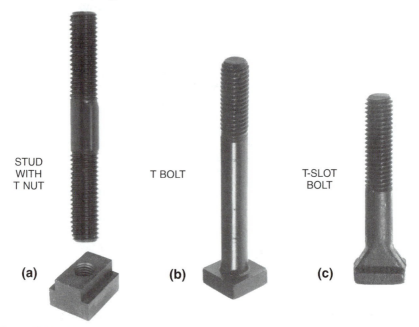

STUD
WITH
T NUT

T BOLT

T-SLOT
BOLT

(a)

(b)

(c)

**Figure 5-26.** *The three most-common center fasteners for clamp straps are studs with T nuts, T bolts, and T-slot bolts.*

*Washers.* Washers, a common item in many workholding applications, are also used with strap clamps. Figure 5-27 shows the common washers for fixturing. These include flat washers, C washers, swing C washers, knurled-face washers, and spherical washers.

The flat washer is one of the more common. With a clamp strap, the main purpose of a flat washer is as a shield between the clamp strap and the fastening element. This prevents any damage to the clamp strap when the fastener is tightened. When installed in a bolted assembly, flat washers are generally needed only under the nut.

The basic C washer is available in two styles, the plain and the swing C washer. These washers work well where a stud and nut, or bolt, clamp a part. With this washer, the nut or bolt is simply loosened a few turns so that the workpiece can be removed without completely removing the nut or bolt. The plain C washer is used where the washer must be completely removed from the assembly.

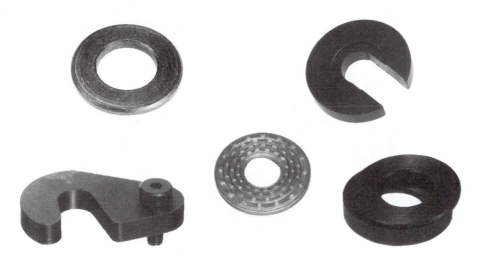

**Figure 5-27.** *Flat washers, C washers, swing C washers, knurled-face washers, and spherical washers are used for many workholding applications.*

Figure 5-28a shows an application with a C washer installed in a groove in a drawbar arrangement. As the cam is rotated, the drawbar tightens against the C washer and clamps the workpiece. The swing C washer is designed for attachment to the workholder and is simply rotated, or swung, out of the way to load or unload parts, Figure 5-28b.

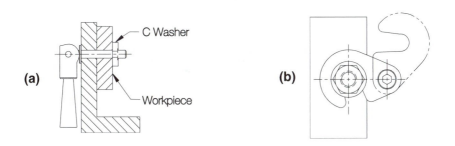

**Figure 5-28.** *The plain C washer is used when the washer must be moved free of the assembly. The swing C washer is attached to the workholder and swung out of the way for loading/unloading.*

Knurled-face washers are flat washers with a serrated face. In the example shown in Figure 5-29, the serrations grip two components of a fixture to prevent any sliding movement. Knurled-face washers are positioned both under the bolt head and nut. A close tolerance on the inside diameter of these washers prevents bolt slippage within the washer.

**Figure 5-29.** *Knurled-face washers are flat washers that have a serrated face to prevent any sliding movement.*

Spherical washer sets are available in two different styles: the first is a set of washers with mating spherical faces, the second is a spherical washer and nut combination, Figure 5-30a. Spherical washers act as

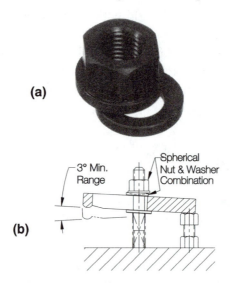

**Figure 5-30.** *Spherical washer sets act as a universal joint between the clamp and the stud or bolt to reduce the effects of bending movement and fatigue on threaded fasteners.*

a universal joint between the clamp and the stud or bolt. These washers reduce the strain and fatigue on threaded fasteners caused by repeated uses on workpieces of varying heights. Even slight differences in part heights can cause considerable fatigue in the fasteners. If left uncontrolled, the fatigue shortens the fastener life and will cause a safety hazard.

The main reasons for fatigue are the variable height of the workpiece and the fixed height of the heel support. Since most parts have some height variation, the fastener is bowed and stressed each time it is clamped against a part. Spherical-washer sets allow limited angular movement of the clamp strap with no effect on the fastener. The spherical joint eliminates the stress on the stud because it compensates for the angular misalignment of the clamp strap and fastener, Figure 5-30b.

*Cam Handles.* Cam handles are made in two general styles, single cams and double cams. These cams often act as the fastening element with strap clamps. Single-cam handles are typically mounted directly on the clamp, as shown in Figure 5-31. Double-cam handles use an eyebolt arrangement, Figure 5-32. The eyebolt acts as the main mount for the double-cam handle.

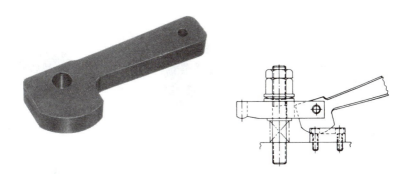

**Figure 5-31.** *Single cams mount at the rear of specially designed clamp straps.*

The major benefit of cam-action clamps is the speed of operation. However, one word of caution: cam clamps rely on the friction between the cam lobe and the clamp or workpiece to maintain clamping force. Some operations with excessive vibration could cause a cam clamp to loosen, due to the inertial force of the handle.

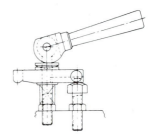

**Figure 5-32.** *A double cam can be used with an eyebolt, instead of a stud and nut, on a slotted-heel clamp strap.*

*Nuts and Knobs.* The nuts and knobs for clamp-strap assemblies are made in a variety of styles. Figure 5-33 shows the nuts most often found

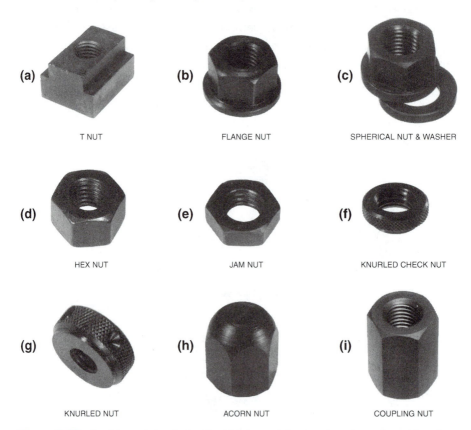

**(a)** T NUT

**(b)** FLANGE NUT

**(c)** SPHERICAL NUT & WASHER

**(d)** HEX NUT

**(e)** JAM NUT

**(f)** KNURLED CHECK NUT

**(g)** KNURLED NUT

**(h)** ACORN NUT

**(i)** COUPLING NUT

**Figure 5-33.** *A wide variety of standard and special-purpose nuts are used for strap clamps.*

in these assemblies. They include the T nut, flange nut, spherical-nut-and-washer set, hex nut, jam nut, knurled check nut, knurled nut, acorn nut, and coupling nut.

The T nuts (a) are often used with flange nuts (b) or spherical-nut-and-washer sets (c) in clamp strap assemblies. T nuts are mounted in the T slots of a machine table and anchor one end of the bolt or stud. The T nut is not threaded all the way through; rather, the threads stop one thread short of the bottom of the nut. This is done to prevent the stud from acting as a jack and breaking the T slot in the table. The flange nut or spherical-nut-and-washer set is used on the other end of the stud. The flange nut, Figure 5-34, combines the advantages of a hex nut and flat washer into a single unit. The spherical-nut-and-washer set goes even further by incorporating the advantages of the flange nut with a universal-joint arrangement, as shown in Figure 5-35.

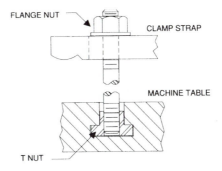

**Figure 5-34.** *Flange nuts are commonly used with studs and T nuts for strap clamps.*

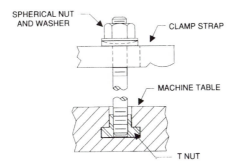

**Figure 5-35.** *Spherical nut and washer sets compensate for workpiece-height variations by permitting some angular movement of the clamp strap.*

Hex nuts and jam nuts are for general-purpose applications. With a first-class-lever arrangement, a hex nut often sets the height of the fulcrum, Figure 5-36. Here the hex nut works with a jam nut (a) to fix the height of the fulcrum. Jam nuts can also be combined with spherical nuts and washers to lock the nuts in a fixed position (b). The jam nut is simply tightened against the other nut, locking both to the stud.

A knurled check nut (c), is often used for the same purpose, but since these nuts are knurled, they are intended for finger tightening. The knurled nut (d), is used when hand tightening is expected. The holes

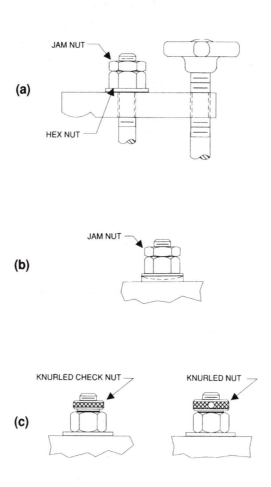

**Figure 5-36.** *Jam nuts are often used with first-class-lever arrangements to set the fulcrum height.*

around these nuts provide for inserting a rod to apply additional clamping force when needed.

Two other nuts are frequently used with clamp straps. The acorn nut is a closed-end nut that protects the ends of studs. These nuts completely enclose the end of a stud to prevent damage to the threads, Figure 5-37. Coupling nuts act as a connector to join two or more studs, Figure 5-38.

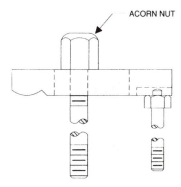

**Figure 5-37.** *Acorn nuts are closed-end nuts used for protecting the ends of studs, and preventing snags.*

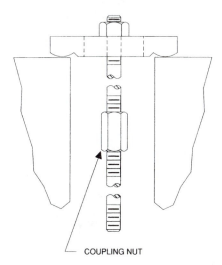

**Figure 5-38.** *Coupling nuts are often used to connect two or more shorter studs to form one long stud.*

In addition to the nuts, knobs are also used with clamp straps. Three common types of heavy-duty knobs are shown in Figure 5-39. They include the palm-grip knob, the hand knob, and the bar knob.

The palm-grip knobs (a) and hand knobs (b) are often used in place of a nut when more speed is needed and reduced holding force is acceptable. These knobs are available as tapped, reamed, or blank knobs, Figure 5-40. The bar knob (c) is used when more force is required. This knob is designed to be tightened with a bar inserted between the four prongs of the knob, Figure 5-41.

**(a)**

**(b)**

**(c)**

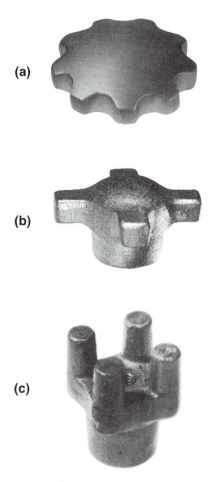

**Figure 5-39.** *Knobs are another element used to apply clamping force to strap clamps.*

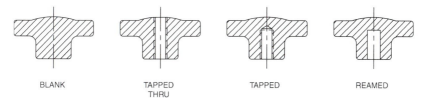

BLANK        TAPPED        TAPPED        REAMED
THRU

**Figure 5-40.** *Hand knobs are available in several styles, including tapped, reamed, and blank.*

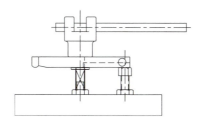

**Figure 5-41.** *Bar knobs can be turned easily by hand, then tightened by inserting a bar between the prongs.*

## Heel Supports

The heel supports for clamp straps provide support at the end opposite the clamping point. The two basic types are threaded and block supports. Threaded supports are mounted either in the clamp strap or in the workholder base; depending on the design of the tool. Block supports can be custom made to suit the clamp height, or standard step blocks can be used.

Common heel supports for strap clamps include clamp rests, clamp rest screws, and stud leveling feet, Figure 5-42.

Clamp rests, the most common, are used with the slotted-heel clamp straps. These supports are normally threaded into the tool body. Once set to the correct height, they are locked in place with a jam nut. The standard clamp rest (a) is made with a hex area below the contact. The miniature stainless type (b) is made with a hole which allows turning with a rod. The clamp rest screw (c) is most often used for drill-

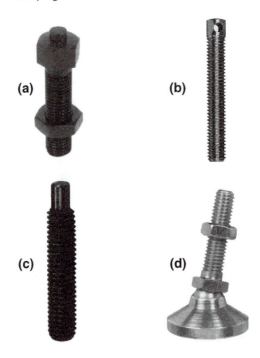

**Figure 5-42.** *Standard types of heel supports used with strap clamps.*

spot-heel clamp straps. These clamp rest screws are also locked at the proper height with a jam nut. Stud leveling feet (d) are normally used with the tapped-heel clamp straps. Figure 5-43 shows how each of these heel supports is used.

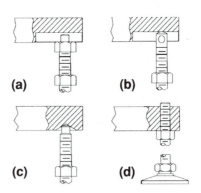

**Figure 5-43.** *Heel supports are selected to meet the requirements of both the clamping operation and the clamp strap.*

Plain-end-type clamp straps most often use step blocks as heel supports. These blocks have a series of serrations, or steps, machined in their mating surfaces, Figure 5-44. The steps in one block engage an identical set of steps in the second block. This design allows the clamp strap to be positioned for different workpiece heights.

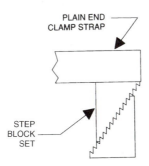

PLAIN END
CLAMP STRAP

STEP
BLOCK
SET

**Figure 5-44.** *Plain-end-type clamp straps often use step blocks as heel supports.*

## Other Clamp-Strap Accessories

Other accessories for strap clamps are shown in Figure 5-45. They include guide blocks (a), clamp springs (b), and finger pins (c). Guide blocks are used as a rear slide block for tapped-heel type clamp straps,

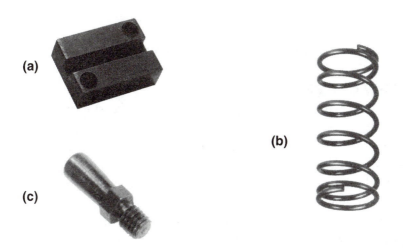

(a)

(b)

(c)

*Figure 5-45.* *Other strap clamp accessories.*

Figure 5-46a. These blocks position the clamping screw and reduce wear on the tool body. Clamp springs, Figure 5-46b, are placed over the stud, between the clamp strap and the tool body. This spring makes loading and unloading the workholder easier because it keeps the clamp strap elevated off of the workpiece. The finger pins, Figure 46c, are occasionally installed in clamp straps to aid sliding movement. Depending on the setup, finger pins can be used on either one or both sides of the clamp strap.

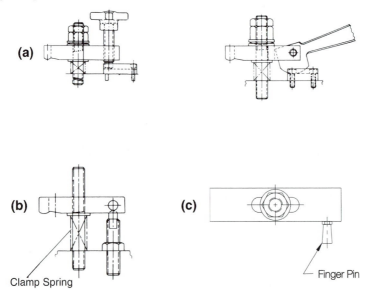

**Figure 5-46.** *Applications of guide blocks, clamp springs, and finger pins.*

## SCREW CLAMPS

Screw clamps are among the simplest and least-expensive clamps in use today. Screw clamps offer the designer more clamping options than many other clamps. For clamping effectiveness and size/force ratio, screw clamps make excellent workholders. But they also have a drawback: their inherently slow clamping speed limits their use in high-production jigs and fixtures.

The principle behind the screw thread is the inclined plane. When applied to a cylinder, the inclined plane converts to a helix. If a thread were unwrapped from a screw, the resulting shape of the thread relative to the screw would be triangular. As shown in Figure 5-47, each part of the triangle has a physical relationship to the screw. The hypotenuse represents the length of the thread around the screw, the side adjacent shows the circumference of the screw, and the side opposite represents the lateral movement of the thread along the screw.

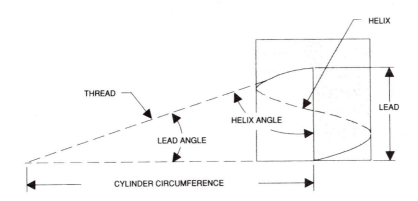

**Figure 5-47.** *A screw thread is an inclined plane. Its features can be explained as they relate to a triangle.*

The helical form of the thread, created by wrapping the triangle around a cylindrical form, allows the thread to transmit lateral movement along the axis of the screw. The specific amount of lateral movement per revolution is called the "lead" of the thread. A term often confused with the lead is the "pitch." The pitch of a thread is the distance from a point on one thread to the same point on the next-adjacent thread. With single-lead threads the lead and pitch are the same, but with multiple-lead threads they are different.

The shorter the lead, the better the holding action, so fine pitches are sometimes selected when vibration might loosen the thread. But with shorter leads, the screw must make more revolutions to cover the necessary line distance. So although the screw thread is an excellent clamping device, the additional time needed to tighten a screw can be objectionable in some situations.

The best and least-expensive solution to the loosening problem is to select standard-size threads and position the screw to minimize the effects of vibration. This can be done with indirect clamping force. As shown in Figure 5-48a, positioning a screw directly on a workpiece

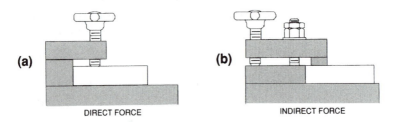

**Figure 5-48.** *Direct and Indirect clamping with a screw clamp.*

increases the chances that vibration will turn the screw. But if the screw applies force to the workpiece indirectly, Figure 5-48b, the vibration is absorbed by the clamp strap before it affects the screw.

Standard clamping screws are designed for either direct- or indirect-pressure clamping. The most-common clamp screws have a hand knob, knurled head, or socket head attached to a threaded stud. Many variations of clamp screws are available for varied needs.

## Hand-Knob Screw Clamps

Hand knobs with screws are a common form of screw clamp. The most-common are shown in Figure 5-49. The hand-knob assembly (a) and the knob-shoe assembly (b) are similar in design. Both are found in either direct or indirect clamping operations, but the hand-knob assembly is usually selected for indirect applications. Here, the dog-point end is often aligned in a slot. The knob-shoe assembly has a swivel pad at the end of the screw, useful in direct-pressure applications. Here, the swivel pad contacts the workpiece and remains stationary while the clamp screw is tightened. This eliminates any part damage that might be caused by the screw rotation, and lessens the effect of vibration.

A variation of the hand-knob assembly is the stud-type four-prong knob (c). This clamp screw has a high-strength plastic handle with a molded-in clamping stud. The knurled-head screw (d) is a smaller variation of the hand-knob assembly and is intended for finger tightening in light clamping applications. This clamp screw also has a dog point, and is made in either steel or stainless steel. The stud-type knurled torque knob (e) is also used for light clamping operations. It is made with a plastic knob and a molded-in clamping stud.

**Figure 5-49.** *Hand knobs with studs are common forms of screw clamps.*

## Swivel Screws

Another form of clamp screw designed for direct-pressure clamping is the swivel screw. The basic styles of swivel screw are shown in Figure 5-50. Swivel screws (a) and (b) are made in either a socket-head or knurled-head style, in either steel or stainless steel.

A complete series of interchangeable feet (c) is available for these clamp screws. The feet allow the swivel screws to be applied to a variety of workpiece shapes. The lefthand thread on the swivel ball (d) allows removing the foot by hand, but prevents backout during clamping.

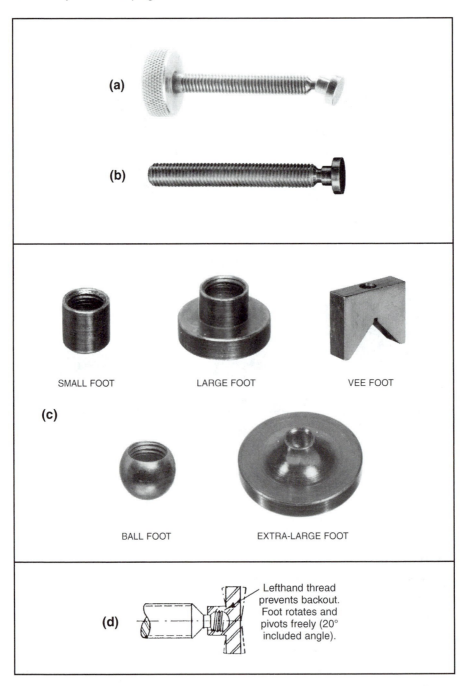

**Figure 5-50.** *Swivel screws are available in a variety of forms with several different contact feet.*

## Adjustable-Torque Screws and Knobs

For applications where clamping force must be controlled, the adjustable-torque thumb screw or adjustable-torque knob, Figure 5-51, can be used. The design of these devices permits controlled torque in the clockwise direction and positive retraction in the counterclockwise direction. The pressure applied is easily adjusted with a hex wrench and can be set to yield 10 to 125 pounds of end force. As shown, the thumb-screw style also accepts a variety of end feet.

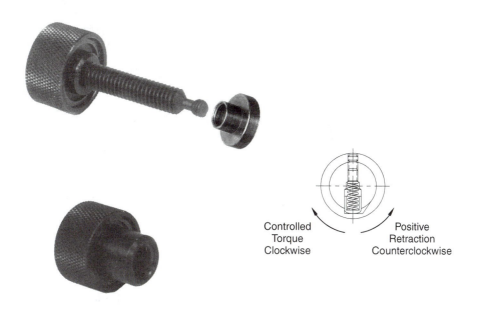

Controlled
Torque
Clockwise

Positive
Retraction
Counterclockwise

**Figure 5-51**. *Adjustable-torque thumb screws or adjustable-torque knobs are for applications where clamping force must be controlled.*

## Quick-Acting Screws and Knobs

One way to increase the speed of a threaded clamp is with a quick-acting screw clamp or knob. Several variations of quick-acting clamps are available. These designs increase the speed of the screw clamp by removing a portion of the threads or by using a special thread form. The Bar-Lok quick-acting screw clamp, Figure 5-52, is an example of this arrangement applied to a clamp screw. The clamping bar has

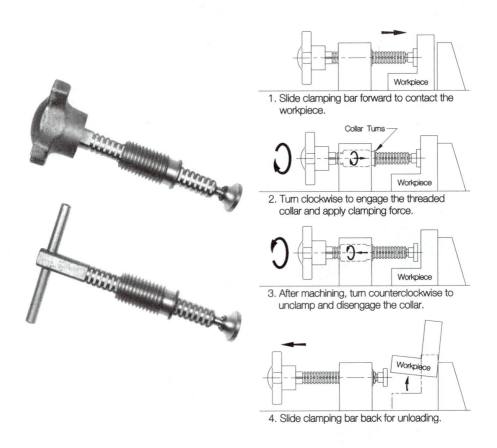

1. Slide clamping bar forward to contact the workpiece.

2. Turn clockwise to engage the threaded collar and apply clamping force.

3. After machining, turn counterclockwise to unclamp and disengage the collar.

4. Slide clamping bar back for unloading.

**Figure 5-52.** *The Bar-Lok quick-acting screw clamp slides into position to increase the speed of the clamping screw.*

a series of locking grooves that engage an externally threaded collar when the clamp is tightened. As shown, the bar is initially retracted to allow loading the workpiece, then advanced to contact the workpiece. Once positioned, the bar is turned to engage the collar and securely clamp the workpiece. The Bar-Lok screw clamp is available with either a knob handle or a sliding handle.

Other variations of the same principle are the quick-acting knobs. As shown in Figure 5-53, these are available as either a knurled knob (a) or four-prong knob (b). The operation of these knobs is shown at (c). A portion of the thread is removed to permit the knob to be tilted

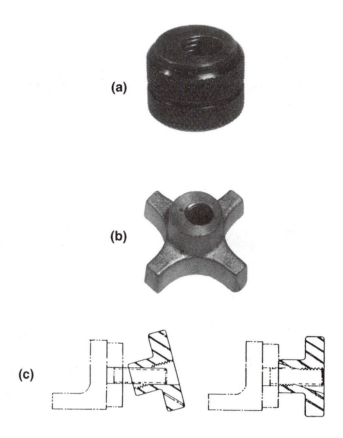

**(a)**

**(b)**

**(c)**

**Figure 5-53.** *Quick-acting knobs are another variation of quick-acting threaded fastener.*

and moved easily and quickly along the thread. Once positioned against the workpiece, the knob is realigned with the threads and turned to tighten.

## Additional Knobs and Handles

Clamping knobs and handles are made in a wide variety of styles and types. Some additional knob styles are shown in Figure 5-54.

The knurled knobs (a) are light-duty finger knobs. The knurled torque knobs (b), fluted knobs (c), four-prong knobs (d), T-handle knobs (e), and star knobs (f) are all high-strength-plastic knobs with a molded-in threaded insert. These knobs are available in many sizes.

**Figure 5-54.** *Knobs are available in many variations.*

For heavier clamping applications, the handles shown in Figure 5-55 may be more appropriate. The handles are made of cast iron or cast aluminum alloy and can be machined for custom applications. This selection of handles includes single and double speed-ball handles (a) and (b), crank handles (c), and hand wheels (d). The handles shown at (e) and (f) are solid and revolving handles, which can be attached to the hand wheels or used to make custom crank handles.

Another handle well suited for many applications is the adjustable handle, Figure 5-56. These handles are available with either a threaded hole (a) or a clamping screw (b). The handle can be set at any angle with an internal spring-loaded spline mechanism (c). This design allows the handle to be moved to any convenient position for clamping.

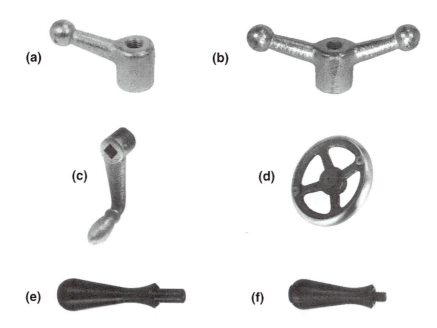

**Figure 5-55.** *For heavier clamping applications, handles may be more appropriate than knobs.*

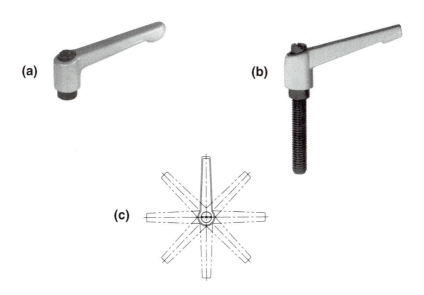

**Figure 5-56.** *The adjustable handle is well suited for restricted spaces because it can be reset to any angle.*

## Latch Screws

Latch screws are another form of screw clamp. They hold other components, such as leaves or swing-away elements, in place. The two common variations are the quarter-turn screw and the half-turn screw, Figure 5-57. The names applied to these screws are descriptive of the action required to position them. Quarter-turn screws are usually used in slots, Figure 5-58a, and turned one-quarter turn (90°) to hold the

QUARTER-TURN
SCREW

HALF-TURN
SCREW

**Figure 5-57.** *The quarter-turn screw and the half-turn screw are latch screws frequently used for workholders.*

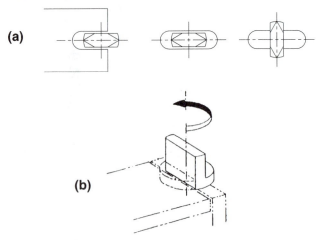

**(a)**

**(b)**

**Figure 5-58.** *Applications for quarter-turn screws and half-turn screws.*

moving element. Half-turn screws are used without machining a slot, as shown in Figure 5-58b. These screws are turned one-half turn (180°) to hold the moving element.

## SWING CLAMPS

Swing clamps are devices which use a swinging clamp arm to speed clamping and unclamping. As shown in Figure 5-59a, swing clamps are made with either knob handles or ball handles in both post-mounted and flange-mounted styles. The post-mounted type (b) is mounted with a single screw in a mounting hole. The flange-mounted

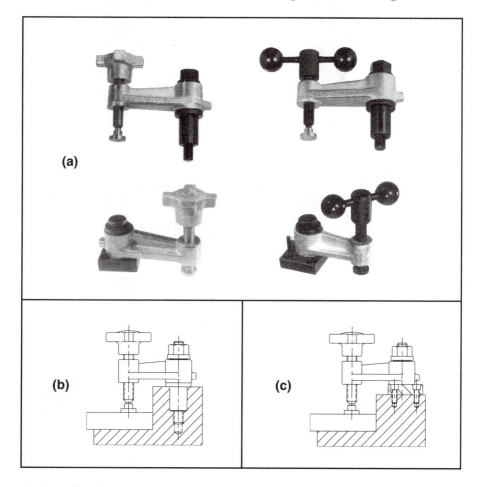

**Figure 5-59.** *Swing clamps use a swing-away arm to increase clamping and unclamping speed.*

style (c) is mounted flat on the tool body and held with cap screws. These clamps also have a swivel foot at the end of the clamp screw. This foot minimizes the effects of vibration and reduces damage to the workpiece.

A smaller variation of the swing-clamp principle is the hook clamp, Figure 5-60. Like swing clamps, hook clamps come in a variety of sizes and two general styles, socket head (a) and hex head (b). Another hook clamp variation is shown at (c). These long-arm hook

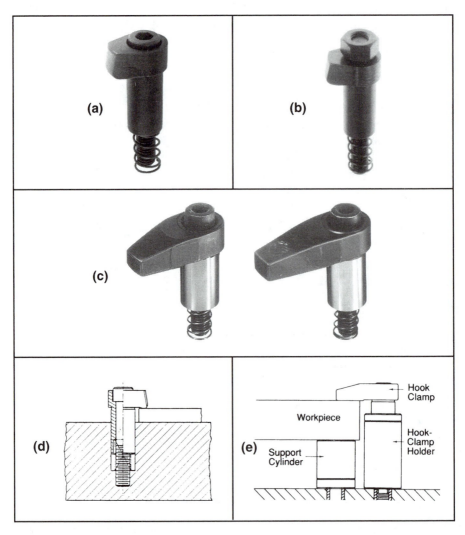

**Figure 5-60.** *Hook-clamp variations.*

clamps are made with either a plain hook or a tapped hook. Hook clamps are mounted either directly in the tool body (d) or in a holder (e).

## EDGE CLAMPS

Edge clamps, as their name implies, are for gripping the side of a part. Like other clamps, edge-gripping clamps are made in a variety of styles. The serrated adjustable clamp, shown in Figure 5-61, is a common form. These clamps transfer the motion of an internal thread into a sliding motion that moves both forward and down along a 45° angle. The mounting slot is slightly angled to prevent movement of the clamp body away from the workpiece under heavy forces. These clamps are

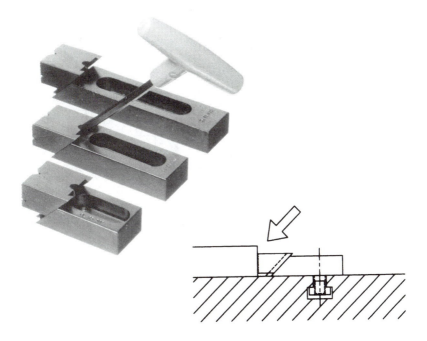

**Figure 5-61.** *Serrated adjustable clamps grip the side of a workpiece by simultaneously pushing forward and down.*

made with either a low nose or high nose. Both types have gripping serrations on the clamping jaws. The high-nose clamp is also furnished with an aluminum cover that prevents damage to soft workpieces.

Another edge clamp, shown in Figure 5-62, is the pivoting edge clamp. Rather than using an angular ramp to apply the holding force, these clamps have a pivot that directs the force. As shown, these clamps are often used with a matching backstop unit to securely hold the work-piece between two jaw elements.

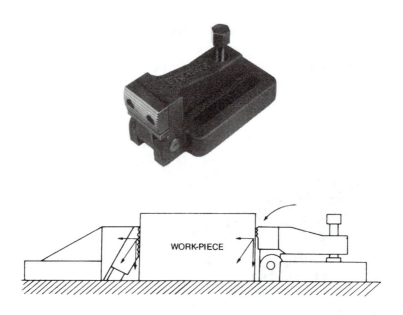

**Figure 5-62.** *Pivoting edge clamps and backstops.*

## C CLAMPS

C clamps are widely used in most types of manufacturing. They are portable clamping devices, but are sometimes used on fixtures, especially in finishing operations. They incorporate a screw clamp into a C-shaped frame. The Duraclamp C clamp, shown in Figure 5-63, is a fiberglass-reinforced thermoplastic clamp designed for the corrosive conditions often encountered in finishing operations. These lightweight C clamps are made in four sizes and have large thumb screws for easier tightening. A torque wrench and socket combination (b) is also available for precise clamping force. The wrenches are preset to 50 inch-pounds of torque, suitable for most clamping operations.

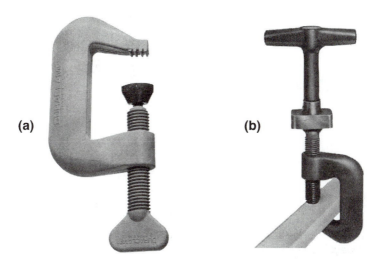

**Figure 5-63.** *Duraclamp C-clamps and accessories. These high-strength plastic C clamps are used in corrosive environments.*

## CAM CLAMPS

Cam clamps are also based on the principle of the inclined plane. The most-common forms of cam clamps are the eccentric cam and spiral cam. Depending on their arrangement, these cams can be used as direct-pressure clamps or indirect-pressure clamps.

## Eccentric Cams

Eccentric cams, as their name implies, apply the clamping force with the action of an eccentric circle. As shown in Figure 5-64, these cams have a mounting hole positioned off center in the cam lobe. The off-center location of the mounting hole produces the "rise," the radial movement of the cam through its clamping cycle.

With eccentric cams, special care must be exercised to keep the cam locked during the clamping cycle. Eccentric cams only have one true locking point. This is the point at which the vertical center lines of the eccentric mounting hole and the lobe are perfectly aligned and

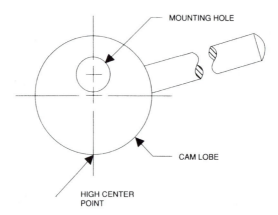

**Figure 5-64.** *Eccentric cams should usually not be used for clamping because they do not provide positive locking.*

exactly perpendicular to the clamped surface. Any other cam position cannot provide a positive lock. For this reason, eccentric cams are generally not suitable for workholding applications.

## Spiral Cams

Continuous-rise spiral cams, though similar in appearance to eccentric cams, differ substantially in their operational principles. Rather than using an eccentric circle to achieve its rise, this cam has an involute curve on the clamping face of the cam. This design provides a wide range of clamping positions instead of just one. The two principal types of spiral cams are the single cam and the double cam, Figure 5-65.

The spiral design, although an improvement over the eccentric design, is still subject to loosening under heavy vibration. For this reason, a mechanical lock is a good idea for applications, such as extremely heavy milling, where considerable vibration can occur.

## Direct-Pressure and Indirect-Pressure Clamping

Cams are typically used in two styles of clamping, direct pressure and indirect pressure. As shown in Figure 5-66, direct-pressure cams apply the force directly to the workpiece (a). Indirect-pressure cams apply the clamping force through a secondary clamping element (b).

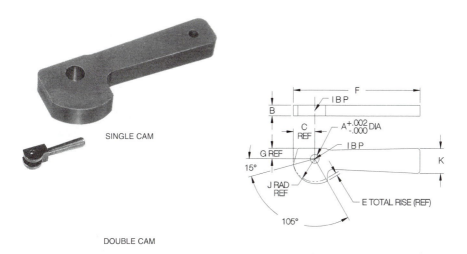

SINGLE CAM

DOUBLE CAM

**Figure 5-65.** *Spiral cams have a continuous rise to lock positively at any point in the clamping range.*

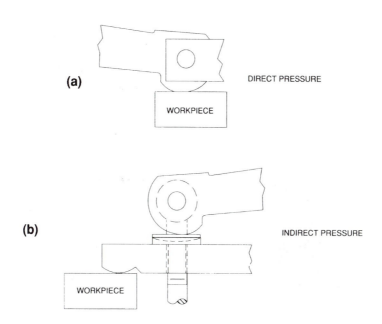

**Figure 5-66.** *Indirect-pressure cam clamps are safer because they are not affected by workpiece vibration.*

The main objections to direct pressure clamps are the susceptibility of the clamps to vibration and the marring of the workpiece. The indirect pressure clamps are less prone to problems, but heavy vibration remains an issue.

The most-suitable applications for cam clamps are where workpiece vibration is minimized. Drill jigs, light-to-medium-duty milling fixtures, inspection fixtures, and assembly tools are all safe, effective applications for cam clamps.

## Commercial Cam-Clamp Variations

Standard cam-clamp designs include cam-action strap clamps, cam edge clamps, up-thrust clamps, and cam-action hold-down clamps. Each of these clamps has been designed to reduce the effects of vibrations and to maximize clamping efficiency.

*Cam-Action Strap Clamps.* Two examples of cam-action strap clamps are shown in Figure 5-67. The clamp shown at (a) has a standard slotted-heel clamp strap and, instead of the stud-and-nut assembly, an eye bolt and double cam. The second clamp, at (b), has a single cam in the cam-type clamp strap. Both these clamps apply indirect force to the workpiece through spiral cams, and are suitable for all but the heaviest machining with excessive vibration.

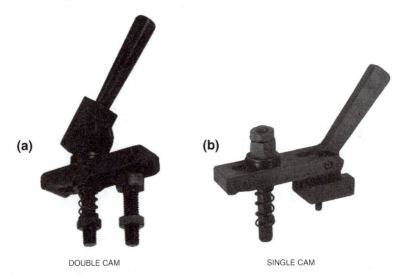

(a)    (b)

DOUBLE CAM    SINGLE CAM

**Figure 5-67.** *Cam-action strap clamps.*

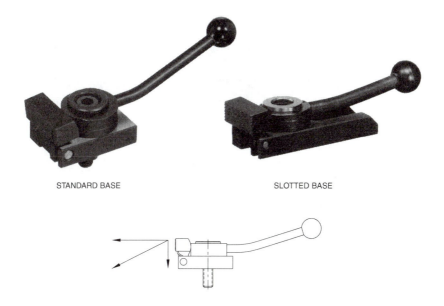

STANDARD BASE            SLOTTED BASE

**Figure 5-68.** *Cam-action edge clamps exert clamping force forward and downward at the same time.*

*Cam Edge Clamps.* The cam edge clamp, shown in Figure 5-68, uses a horizontal single-cam handle with a 180° maximum throw. The pivoting nose element of the clamp applies force both forward and downward to securely hold the workpiece. These clamps are available with either a standard base for fixed-position mounting, or a slotted base for adjustable mounting.

*Up-Thrust Clamps.* The up-thrust clamp, Figure 5-69, is a unique clamp design. It holds a workpiece with pressure from below. As shown at (b), pushing the cam handle down moves the clamping element *upward* against the workpiece. The underside of the top jaw element has a ground surface and acts as a precise locator for the clamped workpiece. This clamp is designed for workpieces that must be located and machined on the top surface.

*Cam-Action Hold-Down Clamps.* The cam-action hold-down clamp, shown in Figure 5-70, is another clamp that uses a cam-action to apply holding force. The cam here is also a spiral design with a 2-1/2° wedge

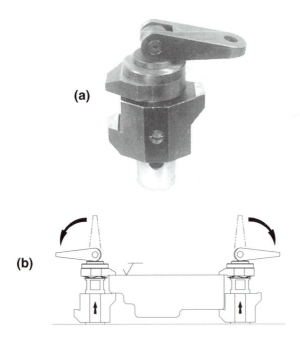

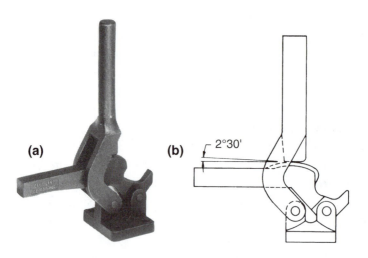

**Figure 5-69.** *Up-thrust clamps apply force by cam action.*

**Figure 5-70.** *Cam-action hold-down clamps use a 2-1/2° spiral cam to achieve a clamping range of 3/16".*

angle. It permits a clamping range of 3/16". This clamp is made of cast steel with flame-hardened wear points. The solid arm can be used to clamp workpieces directly, or tapped to mount a threaded clamp spindle.

## TOGGLE CLAMPS

Toggle clamps have long been a workholding staple in the machine shop. Few clamps offer the overall versatility and efficiency of the toggle clamp. The major benefits of the toggle clamp are an exceptional ratio of holding force to application force, rapid operation, positive locking action, and the ability to be applied in confined areas.

### Toggle-Clamp Operation

Toggle clamps operate using a pivot-and-lever principle referred to as a "toggle action." As shown in Figure 5-71, the basic toggle action is achieved through a system of fixed pivots and levers known as a "four-bar linkage." It provides the clamping action through a series of fixed-length components connected by pivot pins.

When in the unclamped, or released, position, the levers and pivots resemble the arrangement shown at (a). The outer pivot points retract when the clamp is released. This action helps retract or raise the clamp from the workpiece when released. In the advanced position (b), the pivot points are extended. Pressure is exerted on both ends of the linkage to apply the clamping force. In the locked position (c), the levers

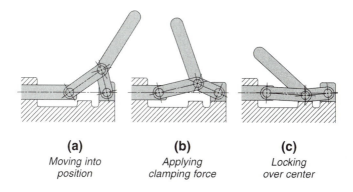

**(a)**
*Moving into
position*

**(b)**
*Applying
clamping force*

**(c)**
*Locking
over center*

**Figure 5-71.** *A toggle action locks positively because one of the pivots moves past the center line of two other pivots, against a stop.*

are set slightly beyond the center point. This position provides the positive locking action of the toggle clamp.

In the clamping process, the action of the clamping handle moves the linkage to its center position and then snaps over the center point to lock the linkage. Although the clamp's maximum force is achieved when the linkage is exactly at its center position, the force lost to moving slightly beyond center is negligible.

The four actions for standard toggle clamps are shown in Figure 5-72. These actions are hold down (a), push/pull (b), latch or pull action (c), and toggle pliers or squeeze action (d). Clamps with the hold-down action are the most common. Push/pull clamps have a straight-line toggle action, capable of applying force in either pushing or pulling directions. Latch-action clamps operate with a pull action.

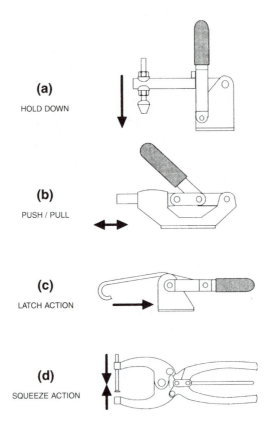

**(a)**

HOLD DOWN

**(b)**

PUSH / PULL

**(c)**

LATCH ACTION

**(d)**

SQUEEZE ACTION

**Figure 5-72.** *The four basic toggle actions are hold-down, push/pull, latch action, and squeeze action.*

Toggle pliers hold parts together with a squeezing action. These clamps work like a toolmaker's clamp or C clamp.

## Hold-Down Clamps

Hold-down toggle clamps are made with either horizontal or vertical handles, Figure 5-73. The horizontal-handle style is clamped by

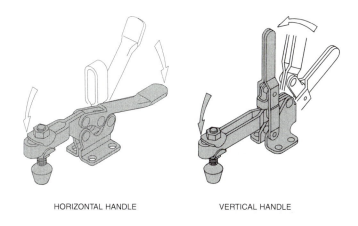

HORIZONTAL HANDLE                    VERTICAL HANDLE

**Figure 5-73.** *Hold-down toggle clamps have either horizontal or vertical handles.*

pushing the handle down. The vertical-handle style is clamped by pulling the handle upward.

As shown in Figure 5-74, these clamps are made with either open or solid clamping arms (a). The open-arm style permits precise adjustments with an adjustable spindle assembly. To get the maximum holding force, the spindle assembly should be positioned close to the handle. For maximum reach, the spindle should be moved to the far end of the clamp arm (b). The solid-arm style is used to directly clamp the workpiece or to attach other contacts. A bolt retainer (c) can also be used with the solid-arm design to mount a spindle or similar contact.

The two standard base styles for these clamps are the flanged base and the straight base, Figure 5-75. The flanged base (a) is used to attach these clamps to horizontal surfaces on the tool body. If a vertical surface is used to mount these clamps, a straight base (b) can be used.

Figure 5-76 shows the wide variety of these hold-down-action toggle clamps available.

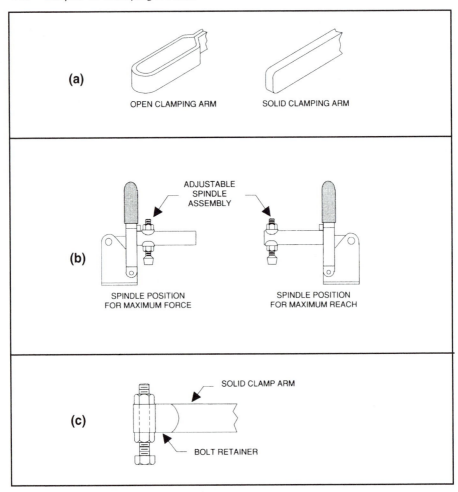

**Figure 5-74.**  *Hold-down toggle clamps have either an open clamping arm for spindle adjustment or a solid clamping arm for fixed spindle position.*

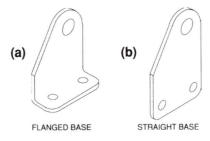

**Figure 5-75.**  *The most-common toggle-clamp base styles are the flanged base and the straight base.*

HORIZONTAL HANDLE

EXTRA-WIDE OPENING

WIDE-OPENING
VERTICAL HANDLE

VERTICAL
T HANDLE

ANGLED-ARM
VERTICAL HANDLE

HEAVY-DUTY
AUTOMOTIVE TYPE

**Figure 5-76.** *Hold-down toggle clamps are available in many styles.*

## Push/Pull Clamps

The push/pull toggle clamps are for applications where a straight-line holding action is required. As shown in Figure 5-77, the toggle action moves and locks the clamp plunger with either a pushing or pulling action. Precise clamping adjustments are made by installing a spindle in the tapped hole at the end of the plunger. Some styles have an externally threaded plunger. The three general types of push/pull clamps are shown in Figure 5-78.

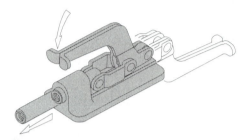

**Figure 5-77.** *Push/pull-style toggle clamps lock the clamp plunger with either a pushing or pulling action.*

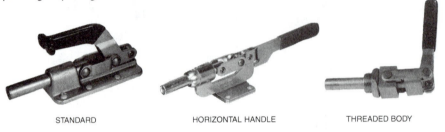

STANDARD                    HORIZONTAL HANDLE                    THREADED BODY

**Figure 5-78.** *Push/pull toggle clamps are available in three general styles.*

## Latch-Action Clamps

Latch-action clamps are designed for applications where a straight pulling action is needed. As shown in Figure 5-79, this clamp's handle is raised to extend the clamping latch. Once the clamping latch is engaged, the handle is moved back to horizontal position to lock the clamp.

These latch-action clamps are made in three styles, shown in Figure 5-80. They are the adjustable-U-bolt (a), adjustable-hook (b), and fixed-length clamps (c). The adjustable-U-bolt and adjustable-hook styles are adjusted by advancing or retracting the clamp element with their threaded bolt ends. The fixed length style in not adjustable. It is intended for applications where adjustment is not necessary. The U-bolt-style clamp is furnished with a matching latch plate to engage the latch bolt. The adjustable-hook style has a precise bend in the hook, to engage a rod or rolled edge.

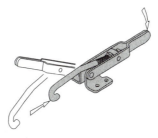

**Figure 5-79.** *Latch-action clamps are for applications where a straight pulling action is required.*

ADJUSTABLE U BOLT      ADJUSTABLE HOOK      FIXED LENGTH

**Figure 5-80.** *The three styles of latch-action clamps.*

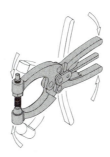

**Figure 5-81.** *Toggle pliers use the toggle action to apply a squeezing motion.*

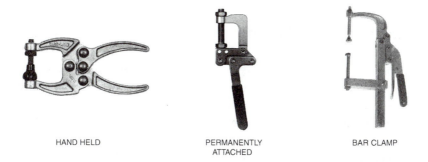

HAND HELD      PERMANENTLY ATTACHED      BAR CLAMP

**Figure 5-82.** *Three styles of toggle pliers.*

## Toggle Pliers

Toggle pliers are a unique design that combines a squeezing motion with a toggle action, Figure 5-81. Hand-held toggle pliers are designed for applications where C clamps were formerly used. The design of the hand-held-style clamp takes either one or two threaded spindles to precisely set its clamping thickness.

Toggle pliers are made in three styles, as shown in Figure 5-82. These are the hand-held (a), permanently attached (b), and bar-clamp (c). Although the hand-held-style toggle plier is not usually attached to a workholder, both the permanently attached and bar-clamp styles can be bolted or welded to the tool body.

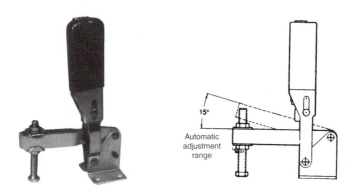

**Figure 5-83.** *Automatic toggle clamps adjust to any workpiece-size variations. Clamping force is exactly the same anywhere within a wide clamping range.*

## Automatic Toggle Clamps

For all its many benefits, the basic toggle action does have one limitation: a limited tolerance for different workpiece heights. Once set to a height, the standard toggle action can accomodate only minor thickness variations. Larger variations require readjusting the clamp spindle. This limitation is inherent in the basic design of the toggle-action arrangement. A standard toggle action has a series of fixed-length components connected by pivot pins. Although excellent for consistent clamping heights, this arrangement does not allow for much height variation.

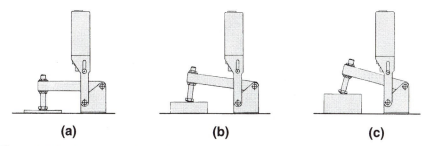

**(a)**          **(b)**          **(c)**

**Figure 5-84.** *An automatic toggle clamp is first set to the average workpiece height. Once set, the clamp automatically adjusts to suit both smaller and larger workpieces.*

An automatic toggle clamp, Figure 5-83, on the other hand, has an internal self-adjusting feature to automatically adjust to different clamping heights. One of the fixed-length components is replaced with a variable-length element in the handle. This allows the clamp to accurately adjust itself for different clamp heights by automatically changing the pivot length. This is done within the handle with a self-adjusting and self-locking wedge arrangement.

The vertical-handle automatic toggle clamp can accommodate variations in clamping heights of up to 15°, for a total automatic-adjusting range of over 1.25" with constant and consistent clamping force. Figure 5-84 shows how the clamp operates. The clamp is first set to the average workpiece height (a). Once set to this height, the clamp automatically adjusts for both smaller (b) and larger (c) workpieces. Clamping force is set initially by turning the screw in the end of the handle.

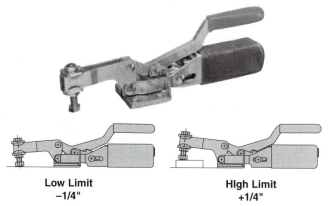

**Low Limit**          **High Limit**
−1/4"          +1/4"

**Figure 5-85.** *Automatic toggle clamps are also available with a horizontal handle for a lower profile.*

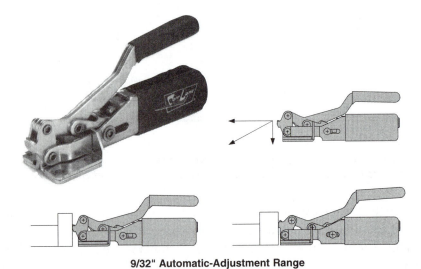

**9/32" Automatic-Adjustment Range**

**Figure 5-86.** *The toggle edge clamp is a self-adjusting clamp that exerts force forward and downward simultaneously.*

Hold-down-type automatic toggle clamps are also available with a horizontal handle, Figure 5-85, for greater overhead clearance. A very unique type of automatic toggle clamp is the toggle edge clamp, Figure 5-86. Like other edge clamps, this clamp applies force both forward and downward.

## Air-Powered Toggle Clamps

Air-powered toggle clamps are similar in design to standard toggle clamps, but they are operated with air cylinders rather than manually. The two styles are shown in Figure 5-87. They are the toggle hold-down clamps (a) and the toggle push clamps (b). The basic toggle

**(a)**                          **(b)**

**Figure 5-87.** *Air-powered toggle clamps are operated with air cylinders, rather than manually.*

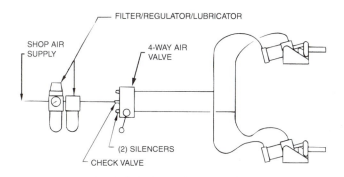

**Figure 5-88.** *A typical plumbing arrangement for air-powered toggle clamps.*

action is the same as that for the manual variation. An example of how these clamps are set up with an air system is shown in Figure 5-88. The shop air supply is directed through a filter/regulator to a four-way air valve which controls actuation of the clamps.

Examples of other air-powered clamping devices without a toggle action are shown in Figure 5-89. They are push cylinders (a), swing clamps (b), and wedge clamps (c). These clamps work well when air pressure is preferred over hydraulic pressure to activate the clamps.

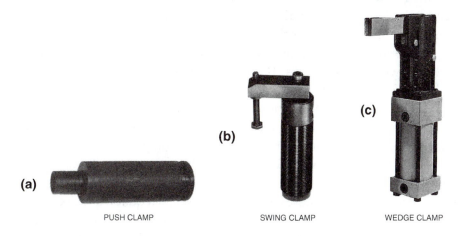

**Figure 5-89.** *Other air-powered clamping devices include push cylinders, swing clamps, and wedge clamps.*

# 6

# POWER WORKHOLDING

Hydraulic and pneumatic clamping have long been a part of manufacturing. Only with the recent introduction of hydraulic systems specifically designed for workholding applications has power clamping emerged as an everyday workholding alternative. This is due, in large part, to the increased demand for faster production and more automation on newer machine tools. But faster production is only one reason for the appeal of power workholding. Today's power-workholding systems offer a wide range of options and capabilities. To make full use of the systems, one must first understand the basic principles of power workholding as well as the range of available components.

Early power-workholding systems, with central hydraulic or pneumatic lines, provided an effective alternative to manual clamping, but left much to be desired. These early systems usually required a major commitment to power clamping, including specialists to design and install plumbing throughout the shop for the hydraulic fluid or compressed air to activate the clamping devices. Not only were the initial costs high, but the systems also added substantially to maintenance costs.

Today, central hydraulic and pneumatic power-workholding systems are still used in some applications, but in most manufacturing industries, small, self-contained hydraulic systems are almost always preferred. Today's hydraulic components are more efficient, and a wide range of hydraulic accessories are available. Hydraulic workholding systems offer two major advantages over pneumatic systems. First, unlike air, hydraulic fluid will not compress significantly, providing more-secure clamping without the springy or spongy action often found with compressed air. In addition, hydraulic workholding components tend to be such smaller than pneumatic components. This is due to the reduced piston size required with hydraulic systems. To generate the same holding forces, pneumatic clamps require approximately 75 times the piston area of hydraulic clamps.

Self-contained power-workholding systems have come a long way since they were first developed. They no longer require design specialists. Instead, the same tool designer who designs the workholder can also completely design the clamping system. The power units are smaller, and can easily be installed on, or next to, the machine tool. With the variety of newer, more-sophisticated components and controls, the cost of power-clamping systems compares favorably with manual clamping.

Throughout the remainder of this chapter, references to "power workholding" describe these newer self-contained hydraulic power-workholding systems.

## ADVANTAGES OF POWER WORKHOLDING

Today, self-contained hydraulic workholding systems offer the designer and machine operator many advantages. The following are important benefits to consider when planning a system. Though covered separately here, in practice the advantages of power workholding are interdependent. Each benefit may actually lead to several other production advantages.

### Faster Clamping

Increased clamping speed is one of the more-obvious advantages of power workholding. Rather than taking several minutes to manually tighten and loosen the clamps on a workholder, a machine operator can activate the complete clamping system from a single point in a matter of seconds. The increased clamping speed thereby reduces the non-productive time which characterizes the loading and unloading cycles. But reduced loading time is only one reason to select power workholding. There are other less-obvious reasons that are even more important.

## Faster Machining

In addition to the reduction of clamping time, power-workholding systems allow much-faster machining cycles. Power clamps offer added security. Instead of relying on the operator to properly tighten the clamps, power-operated clamping systems provide consistent clamping forces. Holding forces can also be adjusted to suit the specific requirements of the workpiece. This permits clamping forces to be increased, allowing heavier feeds and faster speeds.

## Improved Part Quality

Improved part quality is perhaps the greatest benefit of power-workholding systems. These systems improve overall quality and reduce rejected or scrapped parts by providing consistent, controllable clamping forces and self-adjusting work supports.

*Consistent and Repeatable Operation.* A major feature of power-workholding systems is consistent and repeatable clamping forces. Manually operated clamps rely solely on the strength and diligence of the operator. Power clamps, however, are controlled by a power source, so the strength and fatigue level of the operator have no effect on the clamping force.

Control of clamping force increases both the safety and efficiency of the machining operation. Likewise, with swing clamps, extending clamps, or other forms of self-positioning clamps, the position of the clamp on the workpiece is established by the clamp. Once properly set and positioned, power clamps perform the same way, part after part, throughout the production run. The amount of operator interaction is reduced, while consistency and repeatability are enhanced.

*Controlled Clamping Force.* Power-workholding systems are adjustable to provide exactly the right amount of clamping force. When either light or heavy clamping forces are required for a workpiece, the clamping force can be adjusted for those specific conditions. A controlled clamping force is important for parts with varying thicknesses, brittle materials, odd shapes, or similar characteristics. Reduced clamping forces can be applied if the workpiece is delicate or has thin cross sections, as with some cast parts. Conversely, if the workpiece requires greater holding forces, power clamps can also be adjusted for additional pressure.

*Automatically Adjusting Work Supports.* Many workpieces require additional support to prevent deflection or vibration during the machining cycle. In these cases, self-adjusting work supports are quite useful. These supports are placed under the workpiece and either advance to meet the workpiece, or are depressed by loading the part in the fixture. Once the required height is achieved, work supports are locked in position by hydraulic pressure and act as additional fixed locators throughout the machining cycle. At the time of unloading, the work supports return to the free position and are repositioned with the loading and clamping of a new part. So, virtually any differences in the supported surface, such as steps or irregular features, are easily accommodated.

## Other Advantages

Power-workholding systems offer several other advantages over manual clamping. These advantages include remote clamp operation, reduced operator fatigue, automatic sequencing, fixture compactness, and increased machine-tool capacity.

*Remote Clamp Operation.* Most fixturing operations require more than one clamp to hold the workpiece. Of these clamps, some may be hard to reach, and present a safety hazard to the operator. A large workpiece, for example, may require six or more clamps to completely hold the part. Power clamps are typically operated together, from a single point. By simply positioning the clamping valve away from the cutters, the hazard of the operator reaching over the part to tighten a clamp close to a cutter is eliminated. The single remote operating point greatly reduces the time and expense of manually positioning and tightening each clamp, while enhancing operator safety and eliminating hard-to-reach clamps.

*Reduced Operator Fatigue.* Operator fatigue is a design consideration often overlooked. Most operators over-tighten clamps in the morning and under-tighten clamps in the afternoon. Power clamping systems reduce operator fatigue by replacing the strenuous activity of clamping and unclamping manually with the consistent and controllable functioning of hydraulically actuated clamps. This consistency and control results is higher production.

*Automatic Sequencing.* Automatic sequencing is the ability of a power-workholding system to operate clamps and other devices in a specific order. In many clamping situations, this feature is important. Power-workholding systems handle sequencing with one or more sequence valves in the hydraulic circuit. These valves activate the clamps and other devices at the proper time.

To reduce the chance of deformed parts, it may be necessary to activate clamps in a specific order. Self-adjusting work supports often reduce the chance of part deflection. When supports are used, the first operation in the sequence locks the supports under the part. Once the supports are in fixed position against the workpiece, the clamps are brought into contact with the part.

The sequence of operation is also important when a clamp must be moved out of the way during a machining operation. A clamping valve moves the clamp out of the way as the cutter passes. When the cutter is clear, the valve then re-clamps the part. In almost every case, a power sequencing arrangement is much faster and more reliable than its manual counterpart.

*Fixture Compactness.* The high holding force of small power clamps often allows loading more workpieces on a fixture, in many cases, because the clamps can be positioned closer together. One result of this change is increased productivity when the chip-cutting time is lengthened and the percentage of time operator attention is required decreases.

*Increased Machine-Tool Capacity.* The consistent holding force of power clamps allows for faster machining rates. The increased production rates translate into greater capacity, especially important on expensive machine tools.

## DESIGN CONSIDERATIONS

The following design considerations are a general guide to power workholding in a variety of fixturing situations. Although each fixturing condition is different, a few general considerations can be applied to virtually any design with power-operated components.

## Operating Pressure

The maximum operating pressure of most modern hydraulic workholding systems is 7500 psi. Components operate comfortably at this pressure, but 6000 psi is recommended as a standard design pressure for workholding fixtures. As shown in Figure 6-1, a standard operating pressure of 6000 psi offers most of the benefits and efficiencies of the higher pressure with a 25% reserve. The reserve can be used if additional holding force is required after the fixture is built. Pressure can be adjusted either up or down for specific fixturing requirements. Permanently marking each fixture with the chosen operating pressure, for future setups, is strongly recommended.

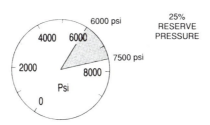

**Figure 6-1**. *A standard design pressure of 6000 psi is recommended for most workholding fixtures, with a maximum pressure of 7500 psi.*

When reduced holding force is required to hold delicate parts, pressure can be reduced. Since lower pressures waste a component's capabilities, reducing the pressure below 6000 psi is not recommended for most applications. Whenever possible, maintaining high pressure while switching to smaller components is preferred. Smaller components require less space and obviously cost less. The minimum recommended operating pressure is generally 1500 psi with an electric power unit and 2200 psi with an air power unit.

## Machining Operations and Fixture Layout

Fixturing designs and workholding concepts are affected by the following factors. To increase the effectiveness and efficiency of the workholder:

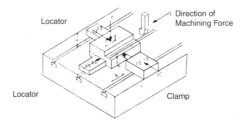

Locator

Direction of
Machining Force

Locator

Locator

Clamp

**Figure 6-2.**

1. Use locators, not the clamping devices, to resist all machining forces. Using smaller, less-expensive clamps is usually possible if the workpiece is correctly located and the locators properly designed.

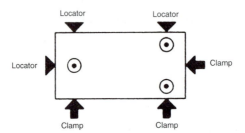

Locator    Locator

Locator    Clamp

Clamp    Clamp

**Figure 6-3.**

2. Select clamping points that are properly supported with positive locators. Position workpiece supports to resist the clamping forces. Unsupported or improperly supported workpieces can distort or deform under clamping force.

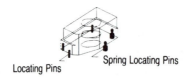

Locating Pins    Spring Locating Pins

**Figure 6-4.**

3. Evaluate the workpiece-loading process and clamping sequence. Loading and clamping can be completely automatic or may need manual steps. In either case, spring-loaded positioners help to properly position the workpiece. Sequence valves are useful to first hydraulically position

and then clamp the workpiece. The total positioning force should be 30-50% of the workpiece's weight to overcome friction.

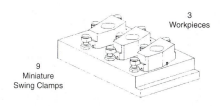

**Figure 6-5.**

4. Determine the number of parts to be loaded on the workholder. Not only will this affect the workpiece positioning and locating methods, but the type of clamps needed as well. For example, loading the three workpieces on the fixture shown in Figure 6-5 requires miniature swing clamps. But, if only a single workpiece is loaded, almost any type or size power clamp could be specified.

## Plumbing Options

The methods to supply hydraulic fluid to the components in a power-workholding system should be carefully studied. Hydraulic fluid is an integral part of the complete workholding system. The supply of fluid should not be viewed as an afterthought. The following work-holder designs show four different plumbing variations for the same fixture, as seen in Figure 6-6.

*1. Tubing Lines on Top of the Fixture Base.*
This is the oldest and most-traditional way to supply fluid to hydraulic clamps. Before manifold mounting, tubing was about the only option available. *Advantages:* Less machining of the fixture-base; reduced building time. *Disadvantages*: Chips are easily trapped around tubing lines; larger fixture-base area required; exposed tubing is subject to damage.

*2. Tubing Lines Underneath the Fixture Base.*
Placing the tubing lines below the working area is an improvement over top-mounted tubing. *Advantages*: No chip traps in the work area caused by the tubing; more freedom to position clamps with tubing out of the way. *Disadvantages*: Larger fixture-base area is still required; more-complicated fixture construction.

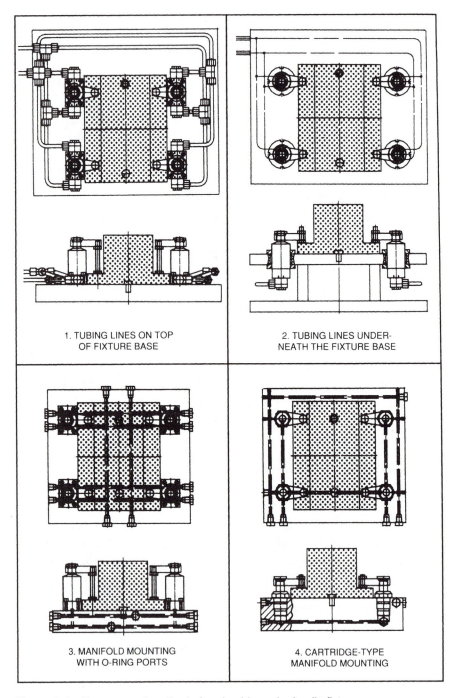

1. TUBING LINES ON TOP
OF FIXTURE BASE

2. TUBING LINES UNDER-
NEATH THE FIXTURE BASE

3. MANIFOLD MOUNTING
WITH O-RING PORTS

4. CARTRIDGE-TYPE
MANIFOLD MOUNTING

**Figure 6-6.** *Four general methods for plumbing a hydraulic fixture.*

*3. Manifold Mounting with O-Ring Ports.*
This option uses passages drilled in the fixture body to feed fluid directly to O-ring ports underneath the clamps. *Advantages*: More-compact fixture; no chip traps in the work area caused by tubing; most-economical construction. *Disadvantages:* Gun drilling is sometimes required for the passages; less freedom to mount clamps in odd positions.

*4. Cartridge-Type Manifold Mounting.*
Similar to mounting with O-ring ports, except clamps are embedded in specially prepared mounting holes. *Advantages*: Most-compact fixture size; great freedom to position clamps in tight places; no chip traps in the working area. *Disadvantages*: Gun drilling is sometimes required; a thicker fixture base is usually required. (This could also be an advantage since it makes the fixture more rigid.)

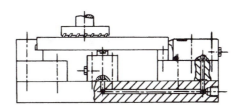

**Figure 6-7.** *Manifold mounting eliminates many tubing problems and allows a single port to feed any number of devices.*

## Manifold Mounting

Manifold mounting, as mentioned above, offers the designer a variety of advantages over tubing. As shown in Figure 6-7, manifold mounting involves drilling one or more holes, usually in the fixture base, to feed fluid to the various components. A single port can feed any number of devices. Installing components in this manner not only eliminates tubing and the problems associated with tubing, but creates a more-compact workholder.

Figure 6-8 shows the steps to prepare the fixture base and install threaded-body components using manifold mounting. When mounting components with O-ring ports, Figure 6-9, grind the mounting surface flat with a 32-microinch, or better, finish. This permits the O-ring to seal properly.

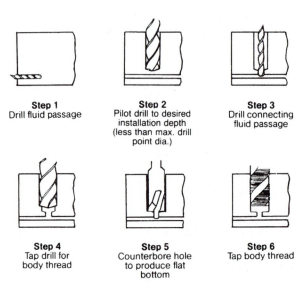

**Step 1**
Drill fluid passage

**Step 2**
Pilot drill to desired installation depth (less than max. drill point dia.)

**Step 3**
Drill connecting fluid passage

**Step 4**
Tap drill for body thread

**Step 5**
Counterbore hole to produce flat bottom

**Step 6**
Tap body thread

**Figure 6-8.** *Steps for installing threaded-body components in a manifold mounting system.*

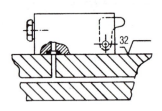

**Figure 6-9.** *Manifold mounting with O-ring ports requires a mounting surface with a 32-microinch finish.*

Additional accessories for manifold mounting are shown in Figure 6-10. The connecting insert shown at (a) connects fluid passages that run through two or more fixture elements. The insert works well when components are mounted on riser elements, pads, or similar attached parts. The expanding plug (b) plugs one end of fluid passages that run completely through the fixture base. The hole end is drilled and reamed larger to insert the plug. The ball element is then driven to the depth shown to properly seal the fluid passage.

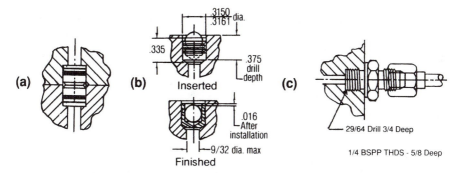

**Figure 6-10.** *Fittings and accessories needed for manifold mounting include connecting inserts, plugs, and port fittings.*

The port fitting (c) is mounted in the entry end of the fluid passage. It connects the hydraulic line from the power source that feeds the workholder. This fitting can be used alone or serve as a mount for quick-disconnect-type fittings. Sealing compounds or tape should never be used when attaching any of these accessories.

## Clamping Force

Too much clamping force can be just as bad as too little. Excessive force causes fixture and machine-table distortion or even damage. Even a small hydraulic clamp can generate tremendous stresses, Figure 6-11.

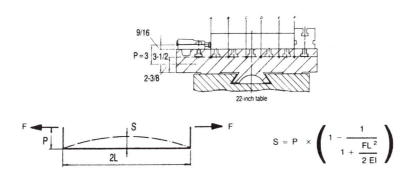

**Figure 6-11.** *Excessive clamping force can cause fixture and machine-table distortion or damage.*

The three 4560-lb edge clamps, positioned as shown, cause the machine-table to bend, or distort. With static beam-bending calculations, the maximum distortion, at point **D**, is about .0006". This small amount of distortion is probably acceptable. But if the clamping point were higher off the machine table, shown by the **P** dimension, the distortion would be much greater. Using higher clamps requires an intermediate fixture plate to increase machine-table rigidity.

## Work Supporting

Work supports, unlike clamps, do not actually exert force on a workpiece. Instead, after adjusting to the location of the workpiece, work supports lock in place and essentially become fixed supports, or rests. The load capacity of a work support increases proportionally as the fluid pressure rises, as shown in Figure 6-12. When selecting work supports, choose sufficient load capacity to resist: (1) machining forces; (2) workpiece weight; (3) clamping forces not resisted by the fixed locators.

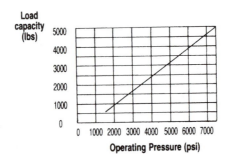

**Figure 6-12.** *The load capacity of work supports increases proportionally as the fluid pressure rises.*

When a work support is positioned directly under a clamp, as shown in Figure 6-13, the load capacity of the work support should be substantially greater than the clamping force. A proper load capacity requires the work support to resist both the static and dynamic clamping loads. These dynamic loads, the repeated "hammering" due to clamping-arm momentum, must not be overlooked. For best results and

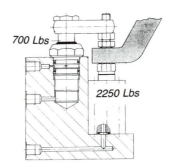

700 Lbs

2250 Lbs

**Figure 6-13.** *When a work support is positioned directly under a clamp, the load capacity of the work support should be at least twice as great as the clamping force.*

for safety, the load capacity of the work support should be at least twice the clamping force.

Another good idea when clamping above a work support is using a sequence valve to activate the work support before the clamp. This prevents the problem of the clamp building up its clamping force faster than the work support builds up load capacity. The sequence valve delays the clamp's activation until the work support has reached sufficient load capacity.

## Single-Acting vs. Double-Acting Clamps

In the majority of power-workholding applications, single-acting clamps should be chosen over double-acting clamps. Double-acting clamping systems are more complicated and expensive. They also require double the plumbing as well as more-complex power sources. There are situations, however, where double-acting systems are better suited or even mandatory. Double-acting systems are needed in the following cases:

1. Fixture designs with moving linkages or clamps that retract heavy loads, Figure 6-14. A double-acting system provides quick, positive return when the weight of the elements is too great for a spring return.

2. Large fixtures with long tubing runs or other flow restrictions, Figure 6-15. Return speed is adversely affected by: (1) pressure drop in tubing

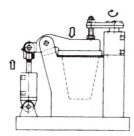

**Figure 6-14.** *Fixtures that move linkages or retract heavy loads should probably use double-acting systems.*

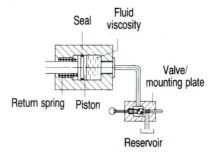

**Figure 6-15.** *Double-acting systems are also well suited for large fixtures with long tubing runs or flow restrictions. Otherwise, return speed is adversely affected by pressure drops in valves and tubing, high fluid viscosity, and frictional forces at piston seals.*

or hoses; (2) pressure drop in valves; (3) high fluid viscosity, especially at lower temperatures; (4) frictional force at piston seals, especially when clamped for an extended time. Extended-time clamping displaces the fluid film on the cylinder walls.

3. Machine-tool interlock. In automated systems where timing and synchronization are important, double-acting clamps are the best choice. By installing pressure switches in both clamping and return lines, as in Figure 6-16, a machine controller knows exact clamp status at all times.

## Position Sensing

　　Clamps with built-in position sensing, such as the swing clamp in Figure 6-17, are ideal for applications where the position of the clamping arm must be monitored. Controlled by a compact, built-in inductive proximity switch, this device ensures that the clamping arm is where it

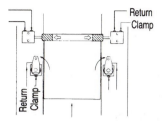

**Figure 6-16.** *Double-acting systems are the best choice for automated systems when timing and synchronization are important.*

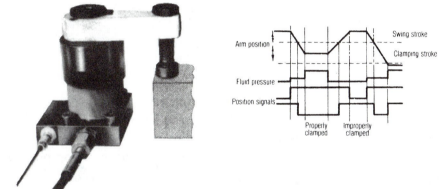

**Figure 6-17.** *Clamps with built-in position sensing provide total closed-loop control of the clamping process.*

is supposed to be at all times. Many other clamps are also available with position sensing.

## Clamping Time

Hydraulic clamping is usually very fast, but it is not instantaneous. To estimate the time required for clamping consider the two phases of clamping shown in Figure 6-18:

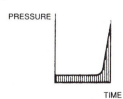

**Figure 6-18.** *Clamping time consists of two phases: extending time under low-pressure free flow, and pressure-building time.*

1. Extending time, under low-pressure free flow.

2. Pressure-building time.

Extending time is fairly easy to calculate. Assuming the fluid required by each clamp and the power unit's flow rate are known, this formula can be used to find the time:

$$
\text{Extending time (sec)} = \frac{\begin{array}{c}\text{Maximum}\\\text{fluid req'd}\\\text{for all clamps}\\\text{(cu. in.)}\end{array} \times \begin{array}{c}\text{Portion of}\\\text{stroke}\\\text{used}\\\text{(usually 1/2)}\end{array} \times 60}{\begin{array}{c}\text{Power Unit}\\\text{flow rate}\\\text{(cu. in./min.)}\end{array}}
$$

One obvious way to reduce clamping time is to set the clamps as close to the workpiece as possible for as little stroke as possible.

After extending, an additional volume of fluid must be pumped into the system to build pressure. This is due primarily to the following factors:

1. Compressibility of the hydraulic fluid (add approximately 4% of the total system volume to build to 7500 psi).

2. Volume expansion of hydraulic hoses (.066 cubic inches per foot).

3. Charging in accumulator, if used.

By calculating the fluid needed for each factor, the pressure-building time can be estimated with the same formula that calculates the extending time. Sequence valves lengthen clamping time since each sequence step requires extending and pressure-building time. To reduce clamping time, set the trigger-pressure of the sequence valve as low as possible. With multiple sequence valves, set the trigger-pressure differences to their minimum allowable values.

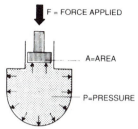

**Figure 6-19**. *Pascal's Law states that pressure applied to a static fluid, completely enclosed, is transmitted equally in all directions.*

## Other Hydraulic Considerations

The basis of hydraulic clamping is Pascal's Law, which states: *if pressure is applied to a completely enclosed static fluid, that pressure will be transmitted equally in all directions,* Figure 6-19. This principle is used to transmit force to remote locations through hoses, tubing, or drilled passages. As shown in Figure 6-20, when hydraulic pressure acts on a clamp's piston area, it generates external force according to the physical relationship:

$$F = P \times A$$

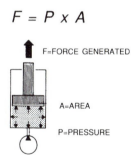

**Figure 6-20**. *When hydraulic pressure acts on a piston, it generates external force according to the physical relationship F = P x A.*

Hydraulically powered clamps can cause strange effects which do not occur with manual clamps. One such phenomenon is fluid shifting between equal-force opposing clamps. As shown in Figure 6-21, two opposed clamps without a check valve will allow the workpiece to float between them. Pushing on one clamp encounters no resistance because, without a check valve, the fluid simply shifts to the opposing clamp. For this reason a pilot-operated check valve should always be installed if equal-force clamps oppose each other.

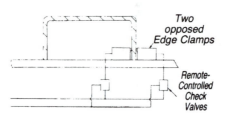

**Figure 6-21**. *Remote-controlled check valves should always be installed if equal-force clamps oppose each other in the clamping operation.*

Another strange effect is the pressure change due to temperature changes in a closed hydraulic system, Figure 6-22. As the temperature changes, so will the pressure. The pressure in a closed hydraulic system changes approximately 80 psi per 1° F. As a safety consideration, excessive temperature changes should be avoided. As an added safety measure, a pressure-relief valve should be installed any time significant temperature changes are anticipated.

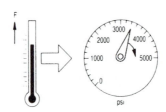

**Figure 6-22**. *The pressure in a closed hydraulic system will change approximately 80 psi per 1° F. A pressure-relief valve should always be installed on any decoupled power-workholding system subject to temperature changes.*

## DOWN-HOLDING CLAMPS

Down-holding clamps, as their name implies, are clamps which apply a downward holding force. The principal clamps in this group are extending clamps, swing clamps, edge clamps, and slideway-locking clamps.

### Extending Clamps

Extending clamps hold a workpiece with a combination of movements. When activated, the clamps first extend straight out, then down against the workpiece. A spring return lifts and retracts the clamping arm when the clamp is released. These clamps can be manifold mounted or plumbed with tubing. Figure 6-23 shows a few applications with extending clamps. The narrow design of these clamps makes them well suited for clamping in narrow recesses or similar places where space is limited. Extending clamps are available in two basic styles.

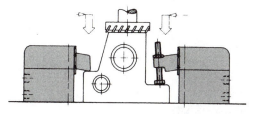

*Choice of two clamping-arm styles. Adjustable arm is more versatile and can be reset after installation, but flush-retracting arm is more compact and offers extra loading clearance.*

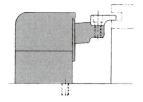

*Clamping in a slot using a custom-made thin-nose clamping tip fastened to the arm with a cap screw. With heavier arm extensions, use a double-acting clamp.*

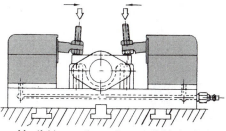

*Manifold-mounting option, available in all Extending Clamp types and sizes, eliminates external plumbing.*

**Figure 6-23.** *Example applications of extending clamps.*

*Flush-Retracting Arm*. The flush-retracting-arm clamp, Figure 6-24, has a clamping arm that, when retracted, sits flush with the face of the clamp body. This design is for applications where additional clearance is needed for loading and unloading workpieces. The retracting arm moves completely clear of the workpiece.

 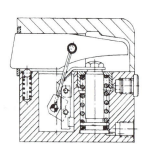

**Figure 6-24.** *Flush-retracting-arm extending clamp.*

*Adjustable Arm*. The adjustable-arm clamp, Figure 6-25, is much the same in design and function as the flush-retracting-arm clamp. However, this clamp has a tapped hole with a contact bolt in the end of the clamp arm. The contact bolt allows vertical adjustment of the contact point.

 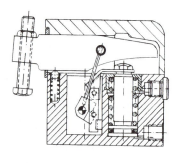

**Figure 6-25.** *Adjustable-clamping-arm extending clamp.*

## Swing Clamps

Swing clamps employ a combination of rotary and linear movements to clamp a workpiece. When activated the clamp arm first swings into position, then clamps down against the workpiece. Rotation is either clockwise or counterclockwise. Although the standard swing angle is 90°, a variety of limited swing angles can be specified when space is limited. As shown in Figure 6-26, the standard options are 0°, 30°, 45°, and 60°.

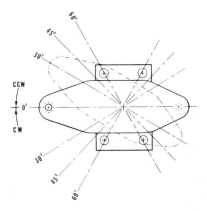

**Figure 6-26.** *90° is the standard swing angle for swing clamps, but other angles are available where space is limited.*

The clamp arm can be mounted to start its 90° swing from any point within 360°, as shown in Figure 6-27. When the clamp is released, the clamp arm lifts and rotates 90° to its original position. This provides ample room for loading and unloading. A unique safety clutch mechanism on the piston rod prevents damage if the clamp arm strikes an object during its swing.

Swing clamps are made in several variations. They come in either single- or double-acting styles with a variety of clamping arms.

*Low Profile.* Low-profile swing clamps, Figure 6-28, are completely self-contained clamping units. They can be mounted on fixture bases, pallets, or directly on machine tables. Their compact design and low height make them well suited for a wide range of applications.

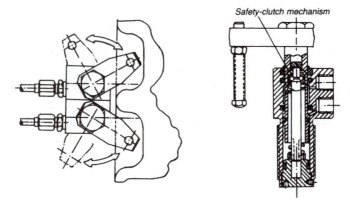

**Figure 6-27.** *A swing clamp's arm can be mounted to start its 90° swing from any point within 360°.*

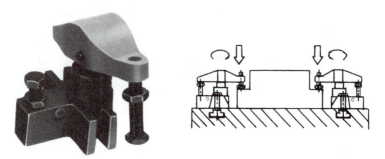

**Figure 6-28.** *Low-profile swing clamps.*

*Flange Base.* Flange-base swing clamps, Figure 6-29, are designed for taller workpieces. The compact base requires a minimum of space on the fixture base. It can be mounted on top of the fixture base or through a fixture plate, as shown. When the standard clamp arm is used,

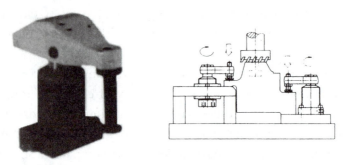

**Figure 6-29.** *Flange-base swing clamps.*

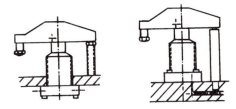

**Figure 6-30**. *Flange-base swing clamps can be fed through standard fittings or manifold mounted.*

Figure 6-30, an additional rear support is also required. The clamps can be either fed though standard fittings or manifold mounted. By eliminating the need for tubing or hoses, manifold mounting further reduces the space needed to mount the clamps.

*All Threaded.* The all-threaded swing clamp, Figure 6-31, is well suited for through-hole mounting. The clamp is positioned to the correct working height and secured in place with two lockrings. Fittings are underneath the clamp to keep plumbing away from the workpiece area. Here again, when the standard clamping arm is used, as shown, an additional rear support is required.

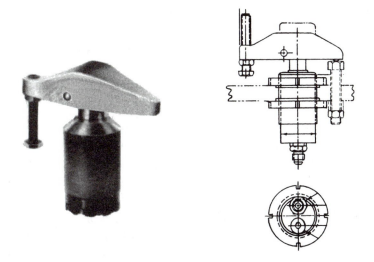

**Figure 6-31.** *An all-threaded swing clamp mounts with two lockrings.*

*Cartridge Style.* The cartridge swing clamp, Figure 6-32, is a threaded-body clamp designed especially for manifold mounting. These clamps are ideal for applications where space is limited and multiple clamps are needed to hold the workpiece. As shown, these clamps are mounted by threading the clamp into the mounting hole up to its body hex.

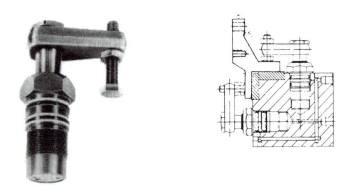

**Figure 6-32.** *The cartridge swing clamp is most-compact type of swing clamp available.*

*Threaded-Body.* The threaded-body swing clamp, Figure 6-33, works well where the clamp must be mounted with the plumbing above the fixture plate. As shown, these clamps can be mounted in a variety of ways.

## Edge Clamps

Edge clamps are ideal when a workpiece is difficult to clamp from above. Edge clamps are usually positioned to take maximum advantage of their low profile. Although the clamp height is low, the clamping action is normally directed high against the side of the workpiece, just below the cutter path. In this application, the edge clamp applies both a lateral and downward clamping force against the workpiece, Figure 6-34. Fluid pressure advances the clamping element against the workpiece. When released, a spring return repositions the clamping element. Two styles of edge clamps are available for a variety of workholding situations.

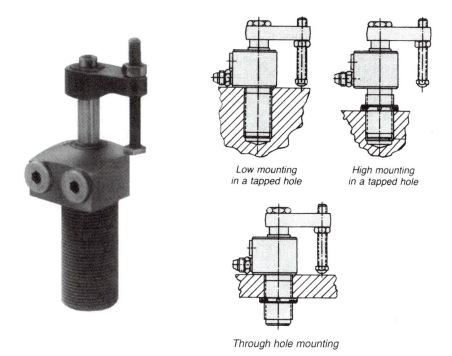

Low mounting
in a tapped hole

High mounting
in a tapped hole

Through hole mounting

**Figure 6-33.** *The threaded-body swing clamp has high-level fittings for exposed plumbing above the fixture plate.*

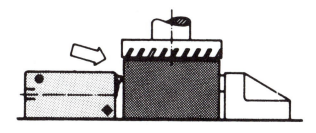

**Figure 6-34.** *Edge clamps apply both a lateral and downward clamping force to the workpiece.*

*Low Block.* The low-block edge clamp, Figure 6-35, applies holding force against the workpiece with a pivoting nose element. When activated, fluid pressure pivots the nose element against the side of the workpiece, applying both lateral and downward clamping forces. These

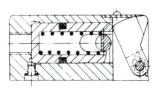

**Figure 6-35.** *The low-block edge clamp has a pivoting nose that clamps just below the cutter path to minimize vibration.*

clamps can be mounted in any position, using four socket-head cap screws, and fed though standard fittings or manifold mounted. Figure 6-36 shows several applications of the low-block edge clamp. This clamp is unique because it sits low yet has a high clamping point, just below the cutter path, for rigid clamping.

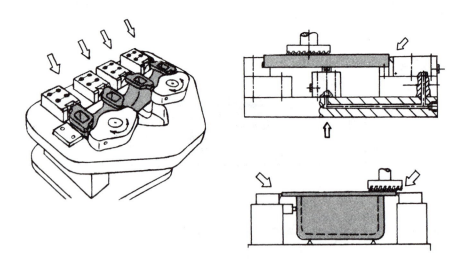

**Figure 6-36.** *Typical applications of the low-block edge clamp.*

*Inclined Plunger.* The inclined-plunger edge clamp, Figure 6-37, advances an inclined plunger against the workpiece with hydraulic pressure. These clamps are normally employed for thin workpieces or in areas where a very-low profile is required. The mounting slots for the clamps provide maximum locational flexibility. They also allow the clamps to be mounted on fixture bases, pallets, or directly on machine tables.

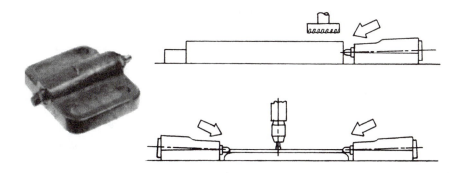

**Figure 6-37.** *The inclined-plunger edge clamp.*

## Slideway-Locking Clamps

Slideway-locking clamps, Figure 6-38, are specifically designed to securely hold machine-tool slideways. If the clamps cannot be tied into an existing hydraulic system, a separate power unit can be used. A machine tool's operation can also be tied to these clamps, so that it will operate only when the clamps have reached a preset locking pressure. These clamps are also well suited for a variety of pallet-locking applications.

## PUSH CLAMPS AND CYLINDERS

Push clamps and cylinders are some of the simplest-yet-most-adaptable hydraulic-clamping devices. Applying fluid pressure provides the basic clamping action of advancing the plunger, and a spring return retracts the plunger. Although a very-uncomplicated design, the combination of their small size, compact design, and variable mounting options makes these clamps well suited for an unlimited number of applications.

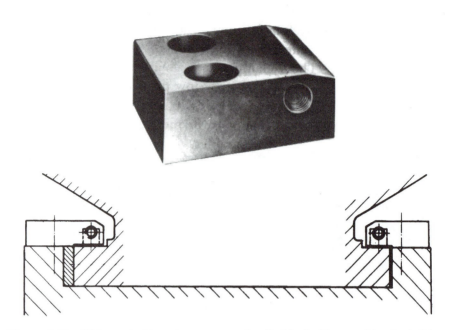

**Figure 6-38.** *Slideway-locking clamps are well suited for locking machine-tool slides.*

## Threaded-Body Push Clamps

Threaded-body push clamps are widely used for a variety of workholding applications. These clamps can be attached to a workholder with accessory brackets or screwed directly into the fixture body, as shown in Figure 6-39. Threaded-body clamps can be fed through

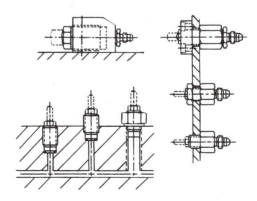

**Figure 6-39.** *Threaded-body push clamps can be manifold mounted, or mounted using adaptors.*

standard fittings or manifold mounted. Their small size permits them to be mounted very close together, less than 5/8" between centers on the smallest models. Their versatility is further extended by a variety of plunger designs and mounting accessories.

*Solid Plunger.* The solid-plunger threaded-body push clamp, Figure 6-40, is the most-basic form of this clamp. As shown in Figure 6-41, these clamps can hold a workpiece directly, or operate levers that clamp the workpiece. In either case, the clamps can be mounted in almost any location.

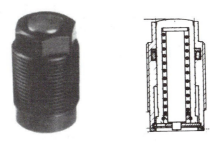

**Figure 6-40.** *The solid-plunger threaded-body push clamp.*

*Tapped Plunger.* The tapped-plunger threaded-body push clamp, Figure 6-42, is a variation of the solid-plunger style. The tapped hole can mount a variety of contact bolts. As shown in Figure 6-43, the standard types of contact bolts are the radius, pointed, and swivel contact bolts. Interchangeable contact bolts permit the clamp to be configured for workpiece requirements.

*Swivel Plunger.* The swivel-plunger threaded-body push clamp, Figure 6-44, is another variation of the solid-plunger-style clamp. Instead of a solid plunger, the clamp has a swivel ball mounted in the plunger. This swivel ball has 9° of movement, which allows the clamp to adjust to workpiece irregularities or slight angles. So, even though the clamped surface may be irregular, the swivel equally distributes the clamping forces.

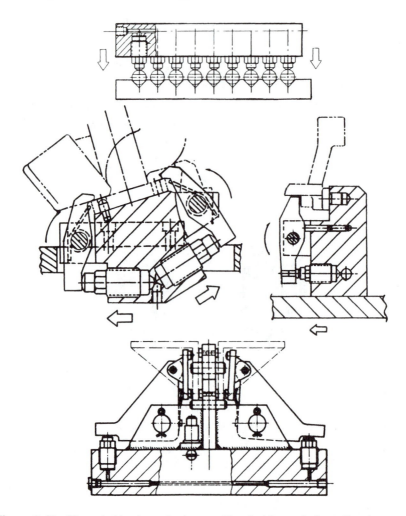

**Figure 6-41.** *Threaded-body push clamps either hold a workpiece directly, or operate secondary clamping elements to clamp the workpiece.*

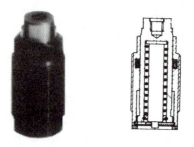

**Figure 6-42.** *The tapped-plunger threaded-body push clamp.*

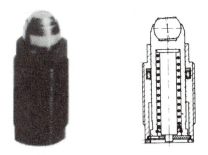

**Figure 6-43.** *Standard contact bolts for push clamps.*
**Figure 6-44.** *The swivel-plunger threaded-body push clamp.*

*Mounting Accessories.* A group of mounting accessories is available for various mounting requirements of threaded-body push clamps, Figure 6-45. The mounting block (a) and mounting flange (b) mount the clamps directly to the fixture body. The vertical adaptor (c) permits the clamp to be mounted vertically and has a mounting slot to allow the adaptor to be positioned precisely. The jam nut (d) securely locks the position of the clamp in a through hole. The feeder cap (e) is used when the clamp is to be directly connected to a hose or tubing.

## Block Clamps

Block clamps, Figure 6-46, are yet-another form of push clamp. Like the threaded-body-style clamps, block clamps can be used for a wide range of workholding applications. Standard block clamps are available in either single-acting or double-acting styles. As shown in

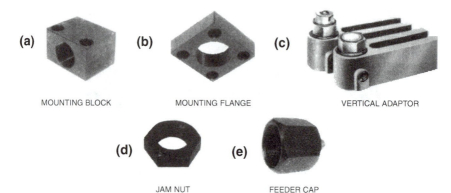

Figure 6-45. *Mounting accessories for threaded-body push clamps.*

Figure 6-47, block clamps can be mounted either vertically or horizontally. Each clamp is furnished with mounting holes but additional mounting holes can be drilled if required. Block clamps can also be plumbed directly or manifold mounted, Figure 6-48. Each block clamp is furnished with a tapped hole in the plunger for a contact button. These contacts can be either custom made, or a standard contact bolt.

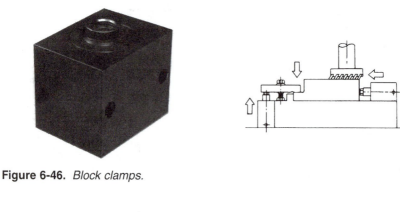

Figure 6-46. *Block clamps.*

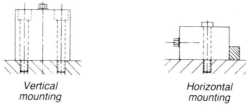

Vertical
mounting

Horizontal
mounting

Figure 6-47. *Block clamps can be mounted vertically or horizontally.*

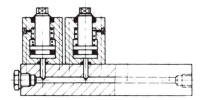

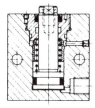

**Figure 6-48.** *Block clamps can be plumbed directly or manifold mounted.*

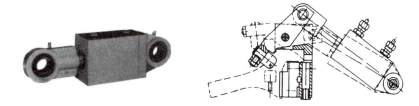

**Figure 6-49.** *Block clamps with bearing eyes.*

## Block Clamps with Bearing Eyes

Block clamps are also available with bearing eyes, Figure 6-49, for operating levers or linkages. These clamps are double-acting and have both a fluid advance and fluid return. The swivel-bearing mounts in each end permit the clamps to be mounted with clevis pins. The swivel feature assures proper alignment throughout the complete clamping/return cycle.

## Threaded-Collar Push Clamps

Threaded-collar push clamps, Figure 6-50, are well suited for mounting though fixture walls or similar workholder elements. In addition, accessory mounting blocks or flanges can also mount these clamps. The clamps also have a tapped hole in the plunger for a contact button. A contact button is always required to prevent wear on the plunger. Either custom-made or standard contact bolts may be used.

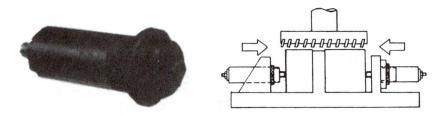

**Figure 6-50**. *Threaded-collar push clamps.*

## PULL CLAMPS

Pull clamps are another variation of clamp. They operate with a pulling action, rather than a pushing action. The two primary types of pull clamps are the hollow-plunger clamp and the block pull clamp.

### Hollow-Plunger Clamps

The hollow-plunger clamp, Figure 6-51, is one of the more-versatile power clamps. Though classified as a pull clamp, these clamps are suited for both pulling and pushing operations. The hollow-plunger design allows a variety of drawbars or fasteners to be used with the clamp. The hollow plunger comes with either a threaded or unthreaded center hole.

One common application of these clamps is converting a manual strap clamp into a power clamp, Figure 6-52. Here the hollow-plunger clamp is mounted at the end of the hold-down stud. The stud can either be threaded into the clamp, or inserted through the hollow plunger of the unthreaded version and held with a nut. A longer hold-down stud is generally required when a hollow-plunger clamp is installed.

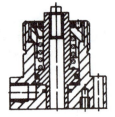

**Figure 6-51**. *The hollow-plunger clamp.*

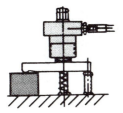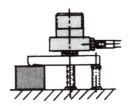

**Figure 6-52.** *Converting manual strap clamps to power operation is one application of the hollow-plunger clamp.*

Figure 6-53 shows additional mounting methods. The first, shown at (a), has four cap screws inserted through the flange and fastened to the fixture body. The clamps can also be mounted with the cap screws inserted through the fixture and threaded into the clamp, shown at (b). As shown at (c), these clamps can also be mounted by their threaded outside diameters. The clamp is installed in a bored hole and held in place with one or two lockrings.

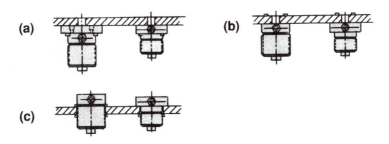

**Figure 6-53.** *The hollow-plunger clamp can be mounted in a variety of different ways, depending on the application.*

The hollow-plunger clamps work well when force must be transmitted to a remote location with a clamping rod or drawbolt. Figure 6-54 shows several applications where hollow-plunger clamps hold a variety of workpieces.

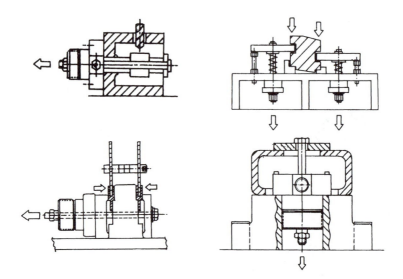

**Figure 6-54**. *Typical applications of the hollow-plunger clamp.*

## Block Pull Clamps

The block pull clamp, Figure 6-55, is another variation of pull clamp. Instead of a hollow-plunger design, however, this clamp has a tapped hole in the plunger. This design permits a variety of standard and custom-made elements to be attached to the clamp. Block pull clamps can be mounted on a flat surface in either a horizontal or vertical position. In addition to acting as a direct-pressure clamp, this clamp can also operate other elements for indirect-pressure clamping setups.

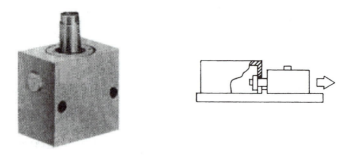

**Figure 6-55**. *The block pull clamp.*

## WORK SUPPORTS

Supporting the workpiece is the primary function of any work-holder. Although solid or manually adjustable supports alone are adequate for some workholders, combining power work supports with solid supports is often a better choice. Hydraulic work supports are sometimes used even with manual clamps. Work supports are available in a variety of types and styles for many applications.

These work supports automatically adjust to the correct workpiece height and, once positioned, securely lock. A sliding-fit pressure sleeve holds the plunger absolutely vertical, even while unlocked. Therefore the plunger remains stationary during the locking process, so there is no loss of accuracy as the support is locked.

Including work supports in a fixture design prevents deflection and vibration caused by the machining forces. As shown in Figure 6-56, work supports act as supplements to the three fixed supports. The locators and supports properly position the workpiece. The work supports

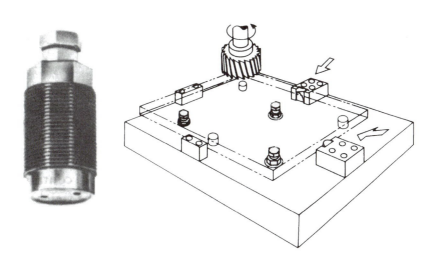

**Figure 6-56.** *Work supports act as supplemental supports in combination with solid locators.*

are then positioned at any points where clamping forces or machining forces can deflect or deform the workpiece. Figure 6-57 shows several fixtures with work supports. Work supports, like many clamps, can be manifold mounted or plumbed with tubing.

The three primary forms of work supports are the spring-extended, fluid-advanced, and air-advanced actions. Each type comes in several shapes and mounting configurations to meet virtually all workpiece-supporting requirements.

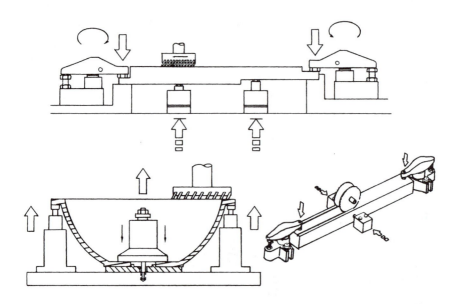

**Figure 6-57**. *Typical examples of work supports applied to workholders.*

## Spring Extended

Spring-extended work supports, Figure 6-58, continuously extend the plunger with a light spring pressure. The weight of the workpiece when loaded depresses the plunger to the proper height. The spring force maintains plunger contact with the workpiece until fluid pressure locks the support in place. Once locked, the work support becomes a fixed, precision-height support.

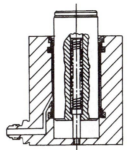

**Figure 6-58**. *A spring-extended work support. The plunger is normally extended by light spring force, until the workpiece's weight depresses it. Applying fluid pressure locks the plunger securely.*

## Fluid Advanced

Fluid-advanced work supports, Figure 6-59, are retracted until fluid pressure is applied, for clear loading. Applying fluid pressure gently advances the plunger. Once workpiece contact is made and a preset resistance is reached, fluid pressure locks the plunger at the precise height. Once locked, the work support becomes a fixed, precision-height support. Note: fluid-advanced work supports should not be used to lift loads.

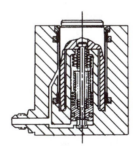

**Figure 6-59.** *A fluid-advanced work support. The plunger is normally retracted for clear loading. Applying fluid pressure gently advances the plunger to the workpiece, then automatically locks it.*

## Air Advanced

Air-advanced work supports, Figure 6-60, advance the plunger with air pressure. These work supports are much like the fluid-advanced type, that is, normally retracted and advanced as air pressure

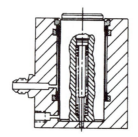

**Figure 6-60.** *An air-advanced work support. The plunger is normally retracted for clear loading, until air pressure advances the plunger to the workpiece (to allow fine tuning contact force). Applying fluid pressure locks the plunger securely.*

is applied. Air-advanced work supports are used for fine tuning the contact force, using a simple air regulator. Contact force can be as light as a few ounces. Leaving air pressure continuously on allows these supports to be used as light-force spring-extended work supports.

## PRECISION VISES

Precision vises are a group of general-purpose power workholders designed to hold many different workpieces. These universal fixtures come in four basics forms: flexible clamping systems, machine vises, self-centering vises, and collet vises.

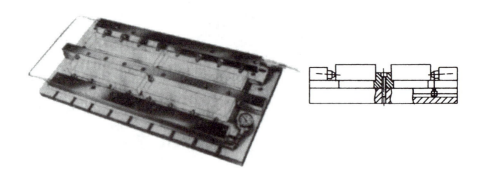

**Figure 6-61.** *Flexible clamping systems can clamp several rectangular parts simultaneously.*

## Flexible Clamping Systems

Flexible clamping systems, Figure 6-61, are precision fixtures well suited for a variety of multiple-part setups. Depending on the system and the size of the workpieces, these systems can hold up to 24 parts simultaneously. Baseplates for these systems have a standard width of 19.50", and are available in four lengths ranging from 15" to 39". These sizes permit the system to hold parts as small as 5/8" square or as large as 13" x 35".

As shown in Figure 6-62, the major components of this system are a fixture plate, manifold blocks, rest plates, and a locating backstop. The fixture base is the baseplate for mounting the other components as well as the workpieces. The manifold blocks hold the threaded-body push clamps that clamp the workpieces. The blocks are plumbed together by simply connecting the fluid passages of the blocks with hydraulic hoses. The rest plates go under the manifold blocks to provide precise positioning of the clamps perpendicular to the baseplate T slots. The locating backstops are attached to the baseplate in the mounting holes. Locating backstops also have side locators to precisely position the workpiece. Figure 6-63 shows how both the manifold blocks

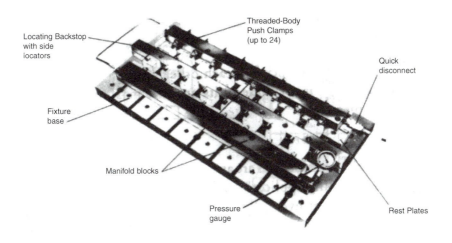

**Figure 6-62.** *The major elements of a flexible clamping system.*

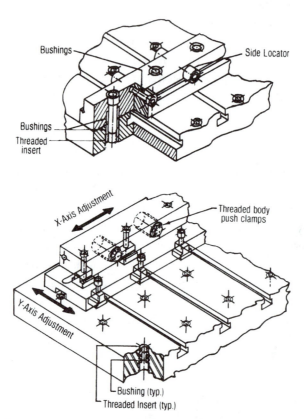

**Figure 6-63**. *Attaching manifold blocks and locating backstops to the baseplate of a flexible clamping system.*

and locating backstops are attached and referenced to each other. Figure 6-64 shows the typical arrangements of these units and the maximum and minimum workpiece sizes.

## Machine Vises

Precision machine vises are another general-purpose workholder for power-clamping applications. As shown in Figure 6-65, these heavy-duty vises come in either hydraulic or hydra-mechanical variations. The hydraulic model has an external power unit to provide the necessary force at the push of a button. The hydra-mechanical vise offers the clamping force consistency of a hydraulic vise, but it does not require an external power source. A graduated collar sets the desired clamping force.

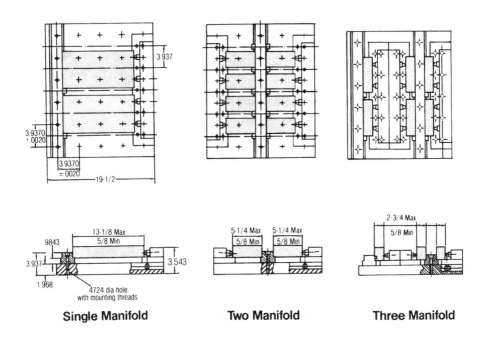

**Single Manifold**  **Two Manifold**  **Three Manifold**

**Figure 6-64**. *Maximum and minimum workpiece sizes that fit in a flexible clamping system.*

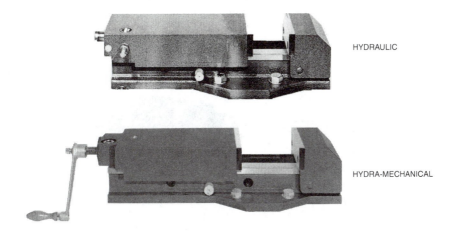

HYDRAULIC

HYDRA-MECHANICAL

**Figure 6-65.** *Hydraulic machine vises clamp automatically at the push of a button. The hydra-mechanical version also provides consistent force using hydraulic power assist, but requires no power source because it is operated by hand.*

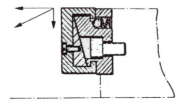

**Figure 6-66.** *Downthrust jaws use a wedge mechanism to apply a downward clamping force to the workpiece as the vise jaws are closed.*

A useful accessory for machine vises is the downthrust jaw, Figure 6-66. The jaws employ a wedge mechanism to apply a downward clamping force to the workpiece as the vise jaws are clamped. Machine vises and vise jaws are covered again more thoroughly in Chapter 9.

## Self-Centering Vises

Self-centering vises, Figure 6-67, combine unique V-shaped jaws with a precision rack-and-pinion mechanism. These vises provide a highly accurate and repeatable clamping action. Workpiece sizes can range from 3/8" to 23" in diameter. These vises have a compact, narrow design that allows them to be positioned virtually anywhere they are required. They are made with two base styles, left foot and right foot,

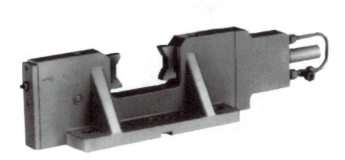

**Figure 6-67.** *Self-centering vises have a precision rack-and-pinion mechanism for accurate centering repeatability.*

Figure 6-68. This leaves the area between the vises completely clear of obstructions.

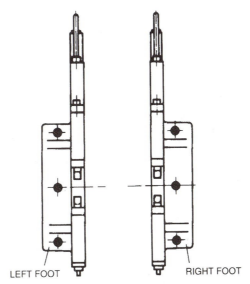

LEFT FOOT RIGHT FOOT

**Figure 6-68**. *Self-centering vises are available with a left-foot and right-foot base.*

## Collet Vises

Collet vises clamp any workpieces that can be held in collets. As shown in Figure 6-69, collet vises are available in both single-collet and triple-collet styles. The vises can be used with any standard or special

**Figure 6-69**. *Collet vises are accurate, versatile fixtures for round, square, and hex-shaped workpieces.*

5C collet. This includes step collets, expanding collets, and emergency collets. Collet vises can be mounted vertically or horizontally, Figure 6-70a. When multiple parts are machined, both the single and triple collet vises can be combined for multiple setups, Figure 6-70b. Both the single and triple collet vises have two single-acting cylinders per collet for clamping, and two return springs per collet to release the collet.

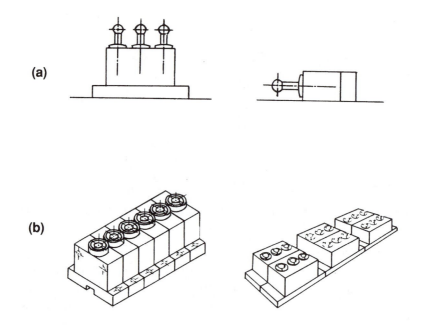

**Figure 6-70.** *Collet vises can be mounted vertically or horizontally, and can be combined for multiple setups.*

## POWER SOURCES

Power sources are the drivers for any power-workholding system. They convert electrical, pneumatic, or manually applied energy into hydraulic power to drive the power-workholding devices. The four primary forms of power sources are electric power units, air power units, hand pumps, and screw pumps. Electric and air power units are usually permanently mounted on a machine tool and used for a variety of different workholders. Hand pumps and screw pumps are usually mounted directly to the workholder.

## Electric Power Units

An electric power unit, Figure 6-71, is a compact, completely self-contained power unit ideal for most power-workholding applications. The units are electrically operated and can be used wherever electrical power is available. The basic power unit includes a pump, reservoir, switches, valves, gauge, and numerous safety devices. The reservoir capacity of a standard power unit is sufficient for virtually all applications.

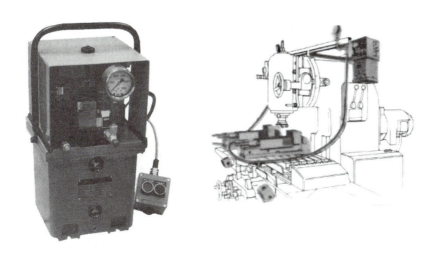

**Figure 6-71.** *Electric power units are a complete hydraulic power source that includes a pump, a reservoir, valves, switches, and a gauge. They are becoming a standard accessory on many machine tools.*

Electric power units are available for either single-acting or double-acting systems, Figure 6-72. With a single-acting system, the electric power unit drives only the clamping cycle. Here a single fluid-supply line goes to the workholder. Spring returns inside the various components force the fluid back into the reservoir. In a double-acting system, the power unit and the workholder have two fluid-supply lines. One controls the clamping cycle; the second controls the return cycle. The clamps, work supports, and other elements of a workholder must be selected based on whether the system is single or double acting.

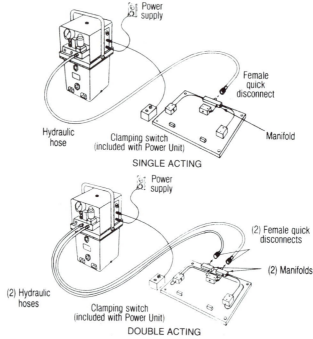

Figure 6-72. *Electric power units are available for both single-acting and double-acting systems.*

## Air Power Units

An air power unit, Figure 6-73, is another compact, completely self-contained power unit. These units are run with normal shop air

**Figure 6-73.** *Air power units are economical hydraulic power sources driven by shop air pressure.*

pressure and can be used wherever shop air is available. Like the electric type, the air power unit includes a pump, reservoir, switches, valves, gauge, and safety devices. The reservoir capacity of a standard air power unit is also sufficient for virtually all applications. Air power units can also operate either single-acting or double-acting systems, Figure 6-74.

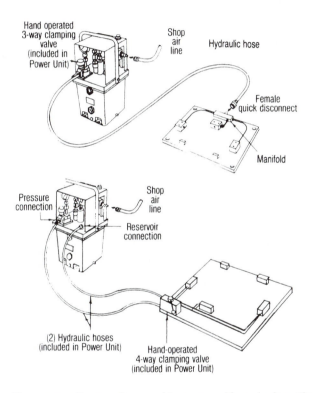

**Figure 6-74.** *Air power units are also used to power either single-acting or double-acting systems.*

## Hand Pumps

Hand pumps, Figure 6-75, are useful for single-purpose workholders when the pump is attached directly to the fixture. Since power is supplied manually to these pumps, neither electricity nor air pressure is needed to operate them. Hand pumps are available as either hand- or foot-operated units. Like the other power sources, hand pumps are completely self contained, Figure 6-76.

**Figure 6-75.** *Hand pumps are compact manual power sources.*

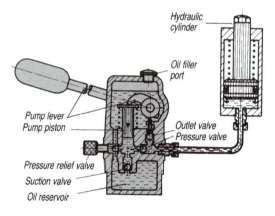

**Figure 6-76.** *Hand pumps are completely self-contained for portability.*

Hand pumps operate with a downward push of the handle. A spring returns the handle to its raised position. Simply repeat the action until the desired pressure is reached. Two-stage hand pumps have an initial high-flow, low-pressure stage for faster clamping. Hand pumps automatically declutch when the preset pressure is reached. To release the pressure in the system, simply lift the handle upward against the spring pressure.

## Screw Pumps

Screw pumps, Figure 6-77, are driven by a rotary motion. The screw pump can be operated manually or with a power torque wrench

**Figure 6-77.** *Screw pumps are small enough to mount directly on a fixture.*

(non-impact type). Like hand pumps, screw pumps are manually operated and neither electricity nor air pressure are required to drive the pump. As shown, screw pumps are available in either a compact block style or in a cartridge/manifold style. Figure 6-78 shows applications with both styles of screw pumps.

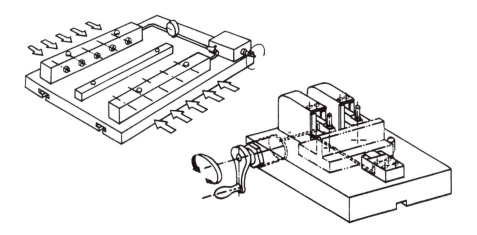

**Figure 6-78.** *Typical applications of screw pumps.*

## VALVES

Valves are an important part of any power-workholding system. Valves control the direction, amount, and timing of fluid flow.

## Three-Way Clamping Valves

Clamping valves are also known as "directional-control" valves because they switch the direction of fluid flow. Three-way clamping valves independently control clamps in a single-acting system. These valves are often found on workholders which hold several workpieces. They permit each clamp to be operated independently of the other clamps. Figure 6-79 shows how the valves control a clamping system.

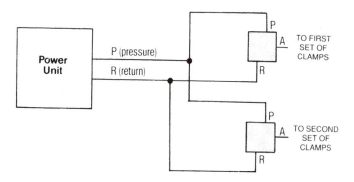

**Figure 6-79.** *Three-way clamping valves are used to operate two or more sets of clamps independently using a single power source.*

Several variations of the three-way clamping valve are available. The function and basic operation of each variation is the same; only the method to activate the valve is different.

*Hand Operated.* The hand-operated three-way clamping valve, Figure 6-80a, is manually operated. Parts are clamped or unclamped by turning the handle 90°. These are the most-common valves used to independently control each clamp in a multiple-clamp system. Individual workpieces can be loaded or unloaded without affecting the other workpieces in the workholder.

*Cam-Roller Operated.* The cam-roller-operated three-way clamping valve, Figure 6-80b, is a mechanically operated valve. Depress the cam roller to unclamp, release again to clamp. Like the hand-operated valve, these valves independently control each clamp in a multiple-clamp system. They are usually part of a mechanically operated automatic system.

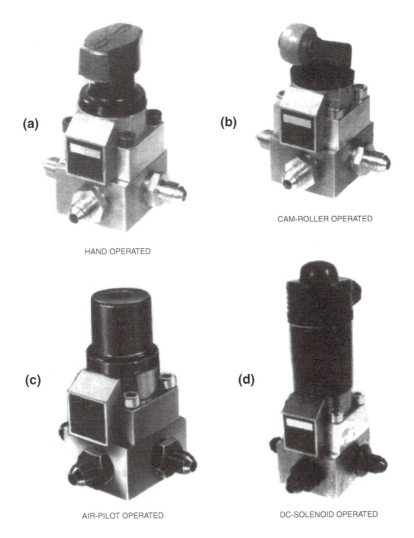

**(a)**

**(b)**

CAM-ROLLER OPERATED

HAND OPERATED

**(c)**

**(d)**

AIR-PILOT OPERATED

DC-SOLENOID OPERATED

**Figure 6-80.** *Various types of three-way clamping valves.*

*Air-Pilot Operated.* The air-pilot-operated three-way clamping valve, Figure 6-80c, is a pneumatically operated valve. Like the other valves, these provide independent control for each clamp in a multiple-clamp system. Air pressure, however, allows remote operation. These valves can also be employed in an automatic system. They are unclamped when air pressure is applied to the valve, then clamped again when air pressure is turned off.

*DC-Solenoid Operated.* The DC-solenoid-operated three-way clamping valve, Figure 6-80d, is an electrically operated valve. Like the other valves, they independently control each clamp in a multiple-clamp system. These valves require a 24-volt DC electric circuit for operation. They are typically found in an automatic clamping system, such as when a machine-tool controller controls the clamping and unclamping cycles. Solenoid valves for 120, 220, and 440 volts AC are also available.

*Clamping-Valve Manifolds.* Instead of mounting each valve independently, a clamping-valve manifold, Figure 6-81, can be specified to simplify the design of the workholder. Manifolds permit a compact arrangement and reduce the required plumbing. When used with valves

**Figure 6-81.** *Clamping-valve manifolds are compact arrangements of three-way valves that reduce fixture plumbing.*

and pressure switches, Figure 6-82, the manifolds can be wired to the machine-tool controller. Such electrical connection ensures proper clamping before the machining cycle begins. Likewise, this setup can also turn off the machine tool if pressure drops below a preset limit during the machining operation.

**Figure 6-82.** *A clamping-valve manifold with valves and pressure switches.*

**Figure 6-83.** *Four-way clamping valves are required for double-acting clamps.*

## Four-Way Clamping Valves

Four-way clamping valves, Figure 6-83, like the three-way type, independently control multiple sets of clamps; however, the four-way valve is for a double-acting system rather than a single-acting system. Figure 6-84 shows how four-way valves control two fixtures with double-acting clamps.

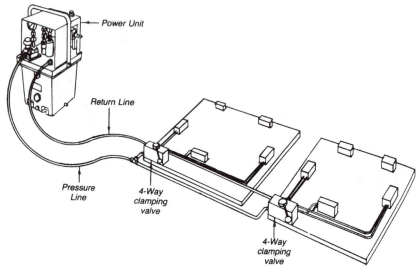

**Figure 6-84.** *Controlling two double-acting fixtures independently with four-way clamping valves.*

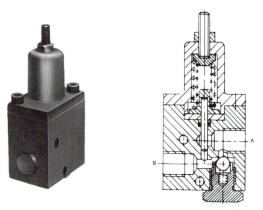

**Figure 6-85.** *Sequence valves automatically sequence locating, clamping, and supporting operations.*

## Sequence Valves

Sequence valves, Figure 6-85, automatically sequence the progression of operations in a power-workholding system. The valves can be used to sequence clamping, positioning, or supporting operations. In Figure 6-86, a sequence valve controls the clamping progression. When the clamps are activated, clamp #1 pushes the workpiece against locator A. Once pressure reaches a preset value of 1500 psi, the sequence

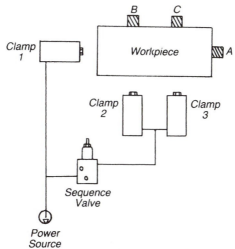

**Figure 6-86.** *Typical application of a sequence valve. Clamp 1 moves into position first, then clamps 2 and 3 are activated.*

valve opens, and clamps #2 and #3 simultaneously push the workpiece against locators B and C. Fluid pressure then rises uniformly in all clamps. Although the example here describes clamp sequencing, the same progression is used for other elements and components as well.

## Pressure-Reducing Valves

Pressure-reducing valves, Figure 6-87, control the pressure in all or part of a clamping system. The valves are quite useful when part of

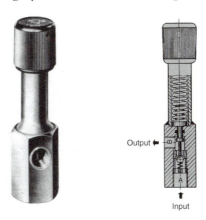

**Figure 6-87.** *Pressure-reducing valves allow having two different operating pressures on the same fixture.*

a system must operate at at a lower pressure. Figure 6-88 shows an example. The work supports operate at standard system pressure, while the swing clamps use less pressure to prevent damage to the workpiece.

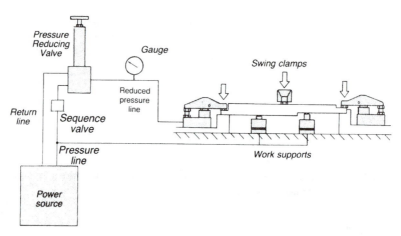

**Figure 6-88.** *Typical application of a pressure-reducing valve. Swing clamps hold a fragile part with light, adjustable force, while work supports hold using full operating pressure.*

## Pilot-Operated Check Valves

Pilot-operated check valves, Figure 6-89, remotely control the flow of fluid in the system. These valves maintain pressure in a clamping system until released by a separate pilot pressure line. In Figure 6-90, the

**Figure 6-89.** *Pilot-operated check valves allow only one-way flow until released by a pilot fluid line.*

valves permit only one-way flow until released. This arrangement provides a safety feature if the line between the power source and fixture becomes damaged. Pilot-operated check valves also work well when

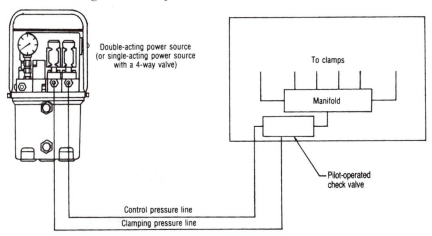

**Figure 6-90.** *Pilot-operated check valves maintain pressure in a clamping system until released by a separate pilot-pressure line. This safety feature keeps the parts clamped if the line between the power source and fixture becomes damaged.*

clamps are positioned with opposing forces, Figure 6-91. The valve allows flow in only one direction, until released, to eliminate any float between the clamps.

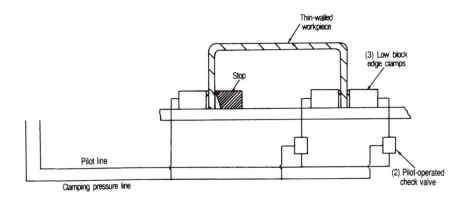

**Figure 6-91.** *Pilot-operated check valves are also used where clamps are positioned with opposing forces. Here the valve eliminates any float between the clamps by restricting the flow.*

## Shutoff Valves

Shutoff valves, Figure 6-92, are used as temporary pressure-holding devices when fixtures are moved with the power source disconnected, or sometimes to shut off unused lines in a multiple-workpiece fixture.

**Figure 6-92.** *Shutoff valves are used to hold pressure in a fixture during transport.*

## Pressure-Relief Valves

Pressure-relief valves, Figure 6-93, are safety devices that protect hydraulic systems from unexpected pressure increases. These increases

**Figure 6-93.** *Pressure-relief valves are a safety device to limit pressure to a set value, such as when fixture temperature increases.*

can be from pressure surges or from temperature changes in closed systems. Any time pressure increases beyond a preset limit, the valve opens and relieves the pressure. Pressure-relief valves are adjustable from 450 to 7500 psi.

## Flow-Control Valves

Flow-control valves, Figure 6-94, restrict the volume of fluid flow from high-flow-rate power sources. The valves restrict the volume of flow in only one direction, into the components. Return flow in the opposite direction is unrestricted.

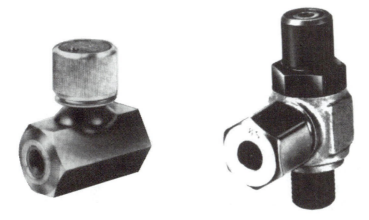

**Figure 6-94.** *Flow-control valves restrict flow in one direction to control clamping speed, while allowing free return flow for unclamping.*

## Check Valves

Check valves, Figure 6-95, permit fluid flow in only one direction, with no return. The arrow marked on the valve indicates the flow direction.

**Figure 6-95.** *Check valves allow fluid to flow in only one direction.*

## FITTINGS AND ACCESSORIES

Fittings and accessories are another important part of every hydraulic workholding system. These components control, direct, and monitor fluid flow throughout the system.

## Rotary Couplings

Rotary couplings allow continuous fluid supply to areas of a workholder that revolve. Without rotary couplings, the range of rotary movements is restricted by the hose; however, when these couplers are part of a design, unlimited rotary motion is possible. Rotary couplings come in several styles and capacities.

*Single Passage.* The single-passage rotary coupling, Figure 6-96, is available in either an elbow or straight configuration. It is used when a single-rotating-line connection is needed. In the example shown, the rotary coupling feeds a fixture mounted on a rotary table from below.

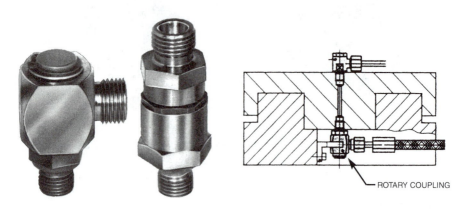

**Figure 6-96.** *Single-passage rotary couplings allow a single hydraulic line to revolve freely.*

*Two Passages.* Two-passage rotary couplings, Figure 6-97a, allow two independent fluid lines to revolve while continuously connected. They are often mounted on top of machining-center pallets to provide a continuous hydraulic connection during machining. Use one passage of the rotary coupling for continuous clamping pressure and the other passage as a return line. Multiple pallet faces can be controlled independently by mounting a separate clamping valve on each

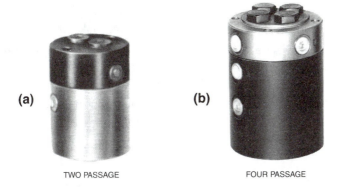

**(a)**                              **(b)**

TWO PASSAGE                    FOUR PASSAGE

**Figure 6-97.** *Multiple-passage rotary couplings are often used to continuously supply fluid on machining-center pallet changers. They are available with two or four passages.*

face (3-way valves for single-acting fixtures, 4-way valves for double-acting fixtures).

*Four Passages.* Four-passage rotary couplings, Figure 6-97b, allow four independent fluid lines to revolve. Figure 6-98 shows how two-passage and four-passage rotary couplings can be used together on a horizontal machining center with a rotary pallet changer. Two passages of the four total passages connect to each pallet.

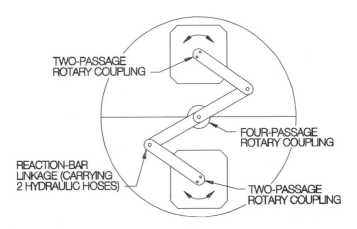

TWO-PASSAGE
ROTARY COUPLING

FOUR-PASSAGE
ROTARY COUPLING

REACTION-BAR
LINKAGE (CARRYING
2 HYDRAULIC HOSES)

TWO-PASSAGE
ROTARY COUPLING

**Figure 6-98.** *Rotary couplings are frequently used on horizontal machining centers with rotary pallet changers. A reaction-bar linkage holds the top portion of each two-passage coupling stationary while the bottom portion revolves freely.*

*Rotary Valve Couplings.* Rotary valve couplings, Figure 6-99, act as both a rotary coupling and a clamping valve. Like rotary couplings, rotary valve couplings work well with multiple fixtures on a rotary platform. But, in addition to supplying fluid, these couplings also have an internal valve that permits them to be used with a load/unload station. One fixture can be unclamped while the other stations remain clamped. Standard rotary valve couplings are available with two through eight stations, for either single- or double-acting clamping circuits.

**Figure 6-99.** *Rotary valve couplings are used for multi-station rotary index tables with an independent loading station.*

## Pallet Decoupling Units

Pallet decoupling units are an alternative to rotary couplings for supplying fluid to machining-center pallets. These compact, all-in-one units allow a fixture to be quickly disconnected from its power source after loading, then remain safely clamped throughout the machining process. Although not as convenient as a rotary coupling, decoupling units are ideal when continuous connection is not possible, as shown in the Figure 6-100 example.

*Pallet Decouplers.* The pallet decoupler, Figure 6-101, is a compact, manually operated coupling unit for individual fixtures. The nitrogen-charged accumulator compensates for pressure variations caused by temperature changes or slight leaks. Pallet decouplers allow the fixture

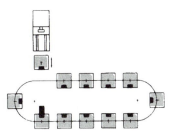

**Figure 6-100.** *Pallet decoupling units are especially useful with pallet pools or in other situations where a rotary coupling cannot be used for continuous fluid connection during machining.*

to be loaded and clamped, then quickly shut off, disconnected, and moved to the machine tool. The unit is reconnected to the power source for unclamping and unloading. Pallet decouplers are available for either single-acting or double-acting fixtures.

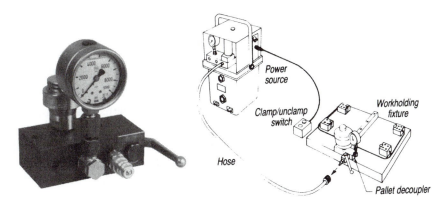

**Figure 6-101.** *A pallet decoupler is a compact unit containing an accumulator, gauge, shutoff valve, pressure-relief valve, and quick disconnect.*

*Tooling-Block Pallet Decouplers.* The tooling-block pallet decoupler, Figure 6-102, is a pallet decoupler for multiple-part clamping. As shown, these decouplers can control four fixtures and work well for two-sided or four-sided tooling blocks. Each station works independently. Individual fixtures can be loaded or unloaded on each of the four sides without affecting the other stations. The built-in accumulator

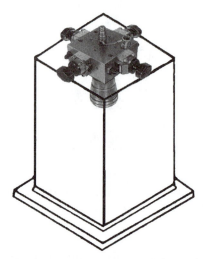

**Figure 6-102.** *Tooling-block pallet decouplers supply fluid to multi-face pallets with one connection. Three-way clamping valves control each face independently. (Similar units are available with a two-passage rotary coupling on top for continuous hydraulic connection during machining.)*

compensates for pressure variations caused by temperature changes. These units allow multi-sided pallets to be loaded and clamped with a single fluid connection.

*Automatic Coupling Systems.* Another coupling method ideal for pallet applications in flexible manufacturing systems is the automatic coupling system, Figure 6-103. This system automates the complete

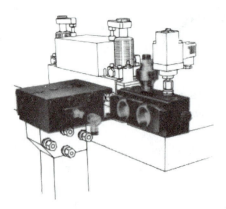

**Figure 6-103.** *Automatic coupling systems consist of a pallet decoupling unit on each fixture, a base unit, and a power unit with a programmable controller.*

clamping/unclamping cycle for either attended or unattended systems. Figure 6-104 shows how the automatic coupling system operates.

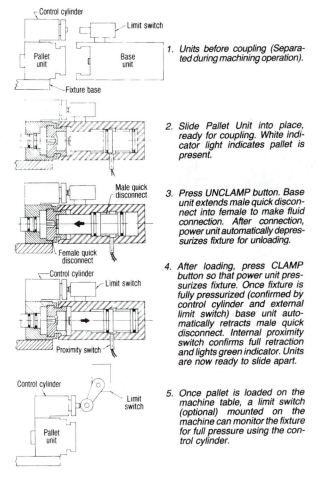

1. Units before coupling (Separated during machining operation).

2. Slide Pallet Unit into place, ready for coupling. White indicator light indicates pallet is present.

3. Press UNCLAMP button. Base unit extends male quick disconnect into female to make fluid connection. After connection, power unit automatically depressurizes fixture for unloading.

4. After loading, press CLAMP button so that power unit pressurizes fixture. Once fixture is fully pressurized (confirmed by control cylinder and external limit switch) base unit automatically retracts male quick disconnect. Internal proximity switch confirms full retraction and lights green indicator. Units are now ready to slide apart.

5. Once pallet is loaded on the machine table, a limit switch (optional) mounted on the machine can monitor the fixture for full pressure using the control cylinder.

**Figure 6-104.** *Automatic coupling systems control the entire pallet-decoupling process automatically.*

## Pressure-Monitoring Devices

One important aspect in power-workholding systems is the ability to monitor system pressure to ensure that everything is correct before the machining cycle begins. The following pressure monitoring devices make absolutely sure that the system is at the proper pressure both before and during the machining cycle.

*Control Cylinders.* Control cylinders, Figure 6-105, monitor system pressure on decoupled pallets. As shown in Figure 6-106, the units are mounted on the pallet and contact a limit switch. These spring-loaded cylinders will extend only when system pressure exceeds the desired setting. If pressure drops, the plunger retracts and will not engage the limit switch.

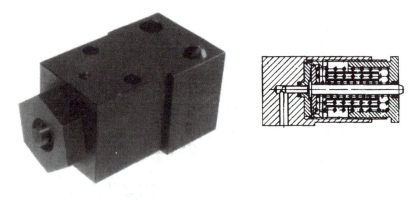

**Figure 6-105.** *Control cylinders extend to indicate correct operating pressure.*

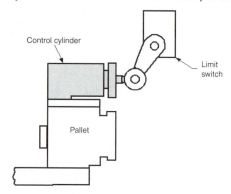

**Figure 6-106.** *Control cylinders are mounted on the pallet and contact a limit switch.*

*Infrared Monitoring.* Another way to monitor system pressure is with an infrared pressure-monitoring system, Figure 6-107. This system is made up of two groups of components, the pallet-mounted components and the base components. The pallet components include a pressure switch, transmitter, and battery pack. The base components include a receiver with preamplifier decoder and delay switch. The pallet components monitor system pressure and emit an infrared-light signal when

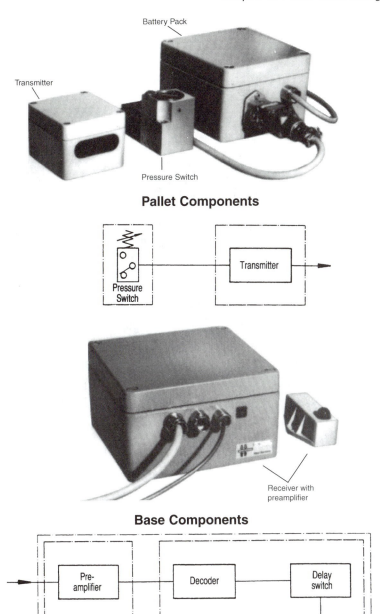

**Pallet Components**

**Base Components**

**Figure 6-107**. *Components of an infrared pressure-monitoring system.*

pressure is above a preset level. The receiver, mounted off the pallet, continuously monitors the signals. If the signal stops, the receiver unit either shuts down the machine tool or sets off an external alarm. Figure 6-108 shows a typical pallet installation.

**Figure 6-108.** *An infrared transmitter, pressure switch, and battery pack sit on top of a palletized tooling-block fixture.*

*Pressure Switches.* Pressure switches, Figure 6-109, are another pressure-monitoring device. These units sense the system pressure, then electrically signal the presence or absence of pressure to the machine-tool controller by opening or closing an electrical circuit. If system pressure drops below a preset value during the machining cycle, the switch shuts down the machine tool.

## Other Accessories

Every power-workholding system needs a variety of items to bring all the components and accessories together to work properly. Although covered last in this chapter, some of these accessories are included in every hydraulic workholder.

**Figure 6-109.** *Pressure switches are safety devices that signal the presence or absence of hydraulic pressure electrically by opening or closing a switch.*

*Manifolds.* Distribution manifolds, Figure 6-110, route the fluid flow on many workholders. These units take a single feed line and split the fluid into several lines. This permits several clamps to be operated from a single input line.

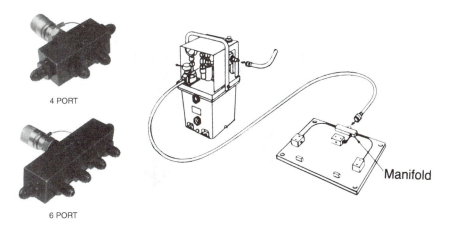

**Figure 6-110.** *Distribution manifolds provide a solid point on the fixture to connect the hydraulic hose from the power unit. From the manifold, fixture hoses or tubing branch out to each clamp.*

*Quick Disconnects.* The quick disconnects, Figure 6-111, are made in two different styles, male and female. They permit the fluid line to be removed from the fixture without special tools or wrenches. The male quick disconnect is normally mounted to the main feed line on the fixture. The female quick disconnect is usually attached to the hose from the power unit.

MALE                                FEMALE

**Figure 6-111.** *Usually a male quick disconnect is mounted on each fixture, and a female quick disconnect on the hose from the power unit.*

*Gauges.* Gauges are another important part of any power-work-holding system. These devices, shown in Figure 6-112, are installed in the hydraulic circuit. They permit the system pressure to be visually monitored and verified at any time.

**Figure 6-112.** *Gauges are an important safety device for every power-workholding fixture.*

*Tubing and Hoses.* The terms tubing and hose are sometimes used synonymously. But normally tubing refers to a rigid fluid conduit and hose refers to a flexible fluid conduit. The tubing and hoses in a power-workholding system are specifically designed and tested for hydraulic-clamping operations. Do not use any other type of hose or tubing for any workholding system.

For permanent applications when a rigid connection between components is desired, tubing should be used. Most tubing used in power-workholding systems is 5/16" O.D. seamless heavy-wall steel tubing with a minimum burst pressure of 22,500 psi (3-to-1 safety factor). Tubing benders, cutters, and flaring equipment are required to install tubing in a hydraulic-workholding system.

Fixture hose, Figure 6-113, is a special small-diameter high-pressure hose. This unique small diameter hose has a braided steel covering to prevent damage to the hose. Because of its tight (1-5/8" minimum) bend radius, fixture hose combines the best features of tubing and standard hydraulic hose.

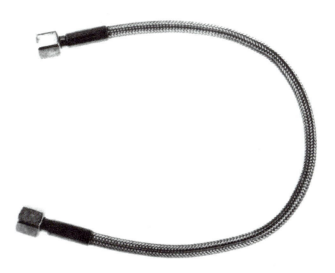

**Figure 6-113.** *Fixture hoses are small-diameter high-pressure hydraulic hoses with a braided steel covering.*

*Fittings.* Fittings, shown in Figure 6-114, are the connectors for hoses and tubing in a power-workholding system. Due to the high operating pressures and the need for leak-free components in power-workholding applications, different hydraulic fittings are required than

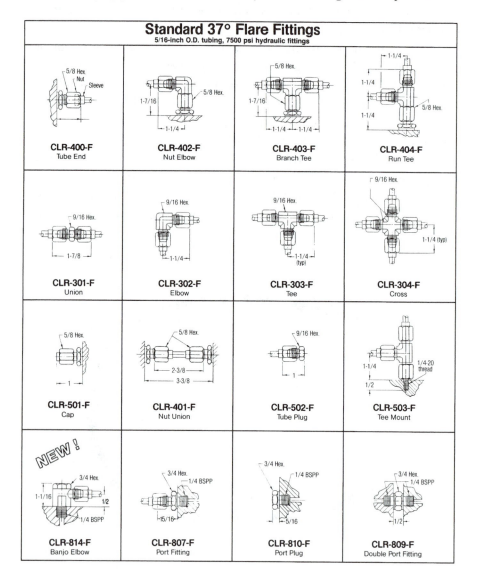

**Figure 6-114.** *A wide variety of fittings are available for hydraulic fixtures. Note: NPT tapered pipe-thread fittings should never be used in a power-workholding system.*

for other industrial applications. NPT tapered pipe threads are never used. Standard fittings are JIC 37° flare or high-pressure compression fittings (5/16" O.D.). All port fittings have straight threads with metal-to-metal sealing, as shown in Figure 6-115.

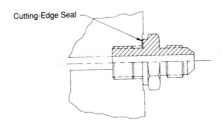

**Figure 6-115.** *All ports in workholding hydraulic systems use BSPP parallel threads with a metal-to-metal cutting-edge seal under the hex.*

# 7

# DRILL BUSHINGS

Drill bushings are a major element in most of today's drill jigs. They act as precision guiding devices for drills, reamers, taps, counterbores, and similar shank-mounted cutting tools. Drill bushings serve three purposes: they locate, guide, and support the cutting tool. Although they serve mainly as guides for cutting tools, drill bushings also have other uses. They work well in assembly tools, inspection tools, and similar devices that require precise alignment and location of cylindrical parts.

The most-common cutting tool for drilling is the twist drill. The design and cutting characteristics of the standard twist drill, although efficient, are not well suited for precision machining. The major reasons are found in the construction of the twist drill.

Twist drills have two angled cutting edges. The cutting edges are usually set 118° apart with a lip clearance angle of approximately 12°. The point formed by these angles is called the "chisel edge." The chisel edge is normally at 135° to the cutting edges of the drill. This design, although highly efficient for cutting, is not effective for centering the tool.

In addition, the material removed to form the flutes and margins of the drill, combined with the standard back taper, greatly reduces the contact area between the twist drill and the hole. The problems of the design increase further because of the unsupported length of the drill. Also, in most production situations, the drill point is not always precisely centered. A drill with an off-center point cuts oversized holes.

When combined, these conditions result in drilled holes that are off center, oversize, out of round, out of alignment, and usually not straight. But simply supporting the twist drill in a drill bushing can greatly reduce, if not eliminate, most of these problems.

## STANDARD DRILL-BUSHING TYPES

Drill bushings come in a wide range of types and styles. The three general categories of drill bushings available are permanent bushings, renewable bushings, and air-feed-drill bushings. As shown in Figure 7-1, drill bushings are identified by letters and numbers. These letters and numbers describe the basic form and specific sizes of each bushing, in a format established by the American National Standards Institute (ANSI). This format consists of one to four letters to identify the bushing type, an OD size in 64ths of an inch, a length size in 16ths of an inch, and the ID of the bushing stated to four decimal positions.

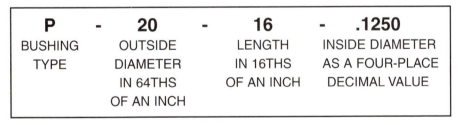

| **P** - | **20** - | **16** - | **.1250** |
|---------|----------|----------|-----------|
| BUSHING TYPE | OUTSIDE DIAMETER IN 64THS OF AN INCH | LENGTH IN 16THS OF AN INCH | INSIDE DIAMETER AS A FOUR-PLACE DECIMAL VALUE |

**Figure 7-1.** *Drill bushings are specified by ANSI letter-and-number designations which identify the bushing type and specific dimensions.*

## Permanent Drill Bushings

Permanent bushings are intended for limited-production applications where bushings are not regularly changed during the service life of the workholder. Permanent bushings are either pressed directly into the jig plate or cast in place. Since these bushings are permanently installed, repeated replacement would cause the mounting hole to wear and reduce the accuracy and soundness of the installation. The following are different varieties of permanent bushings.

*Press Fit.* Press-fit bushings, Figure 7-2, are the most-common and least-expensive permanent bushing. These bushings are identified by the letter P (or PC when the bushing is carbide). Press-fit bushings are designed for one-step operations, such as drilling or reaming. The bushings are pressed directly into the jig plate. They are held in place by the

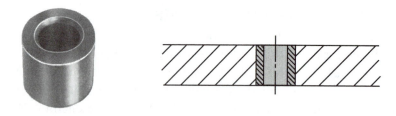

**Figure 7-2.** *Plain press-fit bushings, for permanent installations, are the most-popular and least-expensive drill bushings.*

force of the press fit. Figure 7-3 shows recommended hole size for press-fit bushings. The headless design allows the bushings to be mounted close together and flush with the top of the jig plate. This design, however, offers less resistance to heavy axial loads.

*Head Press Fit.* Head-press-fit bushings, Figure 7-4, are similar to the press-fit bushing in design and application. These bushings, however, are made with a head. Head-press-fit bushings are designed for applications where heavy axial loads might push a press-fit bushing through the mounting hole. The head-press-fit bushings are types H or HC (carbide) bushings. These bushings can be mounted with head exposed, as shown, or counterbored if the bushing must be mounted flush with the top of the jig plate. When the jig plate is counterbored, only the body diameter of the bushing provides the location and only this diameter needs to be reamed. The counterbored area provides clearance for the head and should not be a precision fit. Figure 7-5

| RECOMMENDED HOLE SIZES FOR PRESS FIT BUSHINGS | | |
|---|---|---|
| NOMINAL BUSHING O.D. | ACTUAL BUSHING O.D. | RECOMMMENDED HOLE SIZE |
| 5/32 | .1578-.1575 | .1565-.1570 |
| 3/16 | .1891-.1888 | .1880-.1883 |
| 13/64 | .2046-.2043 | .2037-.2040 |
| 1/4 | .2516-.2513 | .2507-.2510 |
| 5/16 | .3141-.3138 | .3132-.3135 |
| 3/8 | .3766-.3763 | .3757-.3760 |
| 13/32 | .4078-.4075 | .4069-.4072 |
| 7/16 | .4392-.4389 | .4382-.4385 |
| 1/2 | .5017-.5014 | .5007-.5010 |
| 9/16 | .5642-.5639 | .5632-.5635 |
| 5/8 | .6267-.6264 | .6267-.6260 |
| 3/4 | .7518-.7515 | .7507-.7510 |
| 7/8 | .8768-.8765 | .8757-.8760 |
| 1 | 1.0018-1.0015 | 1.0007-1.0010 |
| 1-1/8 | 1.1270-1.1267 | 1.1257-1.1260 |
| 1-1/4 | 1.2520-1.2517 | 1.2507-1.2510 |
| 1-3/8 | 1.3772-1.3768 | 1.3757-1.3760 |
| 1-1/2 | 1.5021-1.5018 | 1.5007-1.5010 |
| 1-3/4 | 1.7523-1.7519 | 1.7507-1.7510 |
| 2-1/4 | 2.2525-2.2521 | 2.2507-2.2510 |
| 2-3/4 | 2.7526-2.7522 | 2.7507-2.7510 |

**Figure 7-3.** *Recommended hole sizes for press-fit bushings in unhardened steel or cast iron jig plates.*

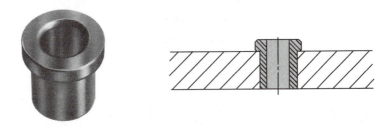

**Figure 7-4.** *Head-type press-fit bushings have a head to resist heavy axial loads.*

shows standard head diameters. The length of the bushing is measured from the underside of the head to the exit end of the bushing.

| BODY<br>DIAMETER | HEAD<br>DIAMETER | HEAD<br>THICKNESS |
|:---:|:---:|:---:|
| 5/32 | 1/4 | 3/32 |
| 13/64 | 19/64 | 3/32 |
| 1/4 | 23/64 | 3/32 |
| 5/16 | 27/64 | 1/8 |
| 3/8 | 1/2 | 3/32 |
| 13/32 | 1/2 | 5/32 |
| 7/16 | 9/16 | 3/32 |
| 1/2 | 39/64 | 7/32 |
| 9/16 | 11/16 | 3/32 |
| 5/8 | 51/64 | 7/32 |
| 3/4 | 59/64 | 7/32 |
| 7/8 | 1-7/64 | 1/4 |
| 1 | 1-15/64 | 5/16 |
| 1-1/4 | 1-1/2 | 1/4 |
| 1-3/8 | 1-39/64 | 3/8 |
| 1-3/4 | 1-63/64 | 3/8 |
| 2-1/4 | 2-31/64 | 3/8 |

**Figure 7-5.** *Dimensions of head-type press-fit bushings.*

*Serrated Press Fit.* Serrated-press-fit bushings, type SP, shown in Figure 7-6, are used for applications where a hardened drill bushing is mounted in a soft jig plate. The bushings have an external mounting surface with both a precision-ground diameter and a serrated, or straight-knurled, area. The ground portion aligns the bushing in the mounting hole in the same way a press-fit bushing does. The serrations prevent any rotational movement from high-torque loads. The serrations also resist axial loads that could push the bushing through the jig plate. These bushings are well suited for jig plates made from aluminum, magnesium, masonite, wood, or similar soft materials.

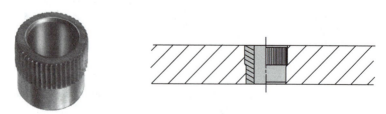

**Figure 7-6.** *Serrated-press-fit bushings have serrations on top to prevent rotation in soft materials, such as aluminum.*

*Serrata Groove.* Serrata-groove bushings, type SG, Figure 7-7, are similar to the serrated-press-fit bushing. They do not, however, combine a precision diameter and serrations. Instead, the serrata-groove bushings are serrated over their entire length. The serrations and grooves cut around the circumference of these bushings suit them for either pressed-in or cast-in-place applications. These bushings offer high torque resistance, but due to their straight-knurled mounting surface, they have a reduced resistance to axial loads. Similarly, since the circumference of these bushings is serrated and not ground, the internal diameter must be used to align the bushing for cast-in-place applications.

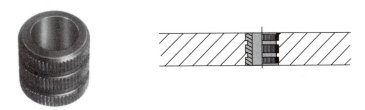

**Figure 7-7.** *Serrata-groove bushings have full-length straight serrations for cast-in-place or potted installations.*

*Diamond Groove.* Diamond-groove bushings, type DG, are another form of bushing for cast-in-place applications. As shown in Figure 7-8, these bushings resemble the serrata-groove bushing, but they have a diamond-pattern knurl rather than straight-pattern knurl on the circumference. The diamond knurl offers high resistance to both rotational and axial forces. Like the serrata-groove bushings, the circumference

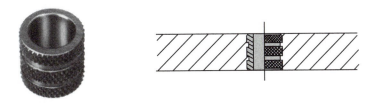

**Figure 7-8.** *Diamond-grove bushings have a diamond-knurled OD for cast-in-place or potted installations subject to heavy axial loads.*

of the diamond-groove bushings are knurled and not ground, so the internal diameter must be used to align the bushing for cast-in-place applications.

Diamond-groove bushings should not be used for press-fit applications. For press-fit applications, straight-serrated bushings are better because when they are pressed into the jig plate, the material displaced by the points of the knurl is moved into the area between the points. A diamond-pattern knurl, on the other hand, will cut the material and actually broach the hole larger.

For cast-in-place applications, the bushings are mounted in holes that have a larger diameter. The space between the outer surface of the bushing and the inside of the hole is filled either with an epoxy resin or a low-melting-point alloy, Figure 7-9.

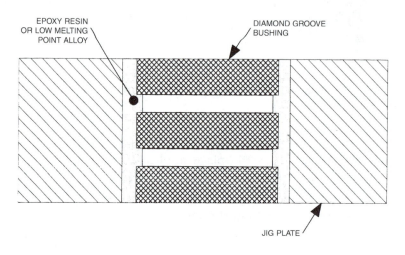

**Figure 7-9.** *Cast-in-place application of the diamond-groove bushing.*

*Template.* Template bushings, type TB, Figure 7-10, are designed for thin jig plates. These bushings permit larger-diameter tools to be used with a thin jig plate. Rather than using a thicker jig plate normally required to support larger-diameter drills. template bushings provide the necessary drill support in jig plates from 1/16" to 3/8" thick. This reduces both the cost and the weight of the jig plate.

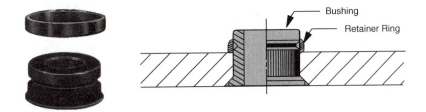

**Figure 7-10.** *Template bushings are for thin template jig plates 1/16" to 3/8" thick.*

Template bushings are installed as shown in Figure 7-11. When locating template bushings, follow the minimum edge distances and hole spacings shown in Figure 7-12a. Once properly located, the mounting hole is drilled and reamed .001" to .003" larger than the mounting diameter of the bushing. The hole is countersunk on the workpiece side to permit the bushing to seat .015" below the surface, Figure 7-12b. The bushing is then inserted and pressed into the hole.

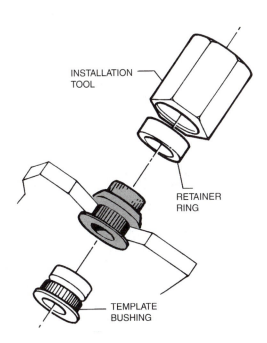

**Figure 7-11.** *Template bushings are installed with an installation tool.*

## 1. LAY OUT HOLES

**(a)**

When laying out holes, observe the minimum hole spacing and edge distance listed below:

| BUSHING OD | A MINIMUM | B MINIMUM |
|---|---|---|
| 3/8 | .60 | .250 |
| 1/2 | .73 | .312 |
| 3/4 | .98 | .438 |

## 2. REAM AND COUNTERSINK

**(b)**

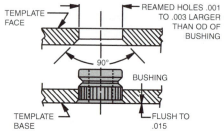

Ream hole .001 to .003 larger than bushing OD. Countersink reamed hole to allow the bushing to seat .015 below flush with surface. For best results, use a piloted countersink tool so that the countersink is concentric and free of chatter marks.

## 3. INSTALL

**(c)**

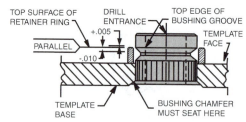

Install Retainer Ring with an arbor press whenever possible. The Installation Tool is also tapped for mounting on a rivet gun or other impact tool. Check that Retainer Ring top surface is within +.005 / -.010 of the bushing groove's top edge before installing.

**Figure 7-12.** *Installation procedure for template bushings.*

The retaining ring is mounted with the installation tool. When mounting the retaining ring, make sure the top of the ring is within +.005"/-.010" of the top of the groove in the bushing, Figure 7-12c, before using the installation tool. The serrations on the bushing circumference prevent rotational movement. The retaining ring both locks the bushing in the jig plate and restricts any axial movement.

*Circuit Board.* Circuit-board bushings, types CB and CBC (carbide), Figure 7-13, are available in either headless or head-type styles. These bushings are specifically designed for small drill sizes. Circuit-

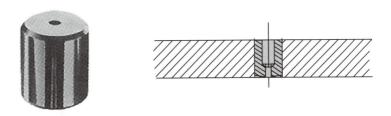

**Figure 7-13.** *Circuit-board bushings are designed to accommodate the large shanks of circuit-board drills.*

board bushings are available for drill sizes from #80 to 9/64" and are made in a variety of styles for specific circuit-board-drilling machines. Figure 7-14 shows some of the more-common forms of circuit-board bushings.

## Renewable Drill Bushings

Renewable bushings are designed for applications where the bushings must be changed regularly during the service life of the workholder. Bushing changes are made when bushings wear out, or when multiple operations are performed in the same hole. With multiple operations, two or more drill bushings are used to produce the desired hole. The two principal forms of bushings for renewable installations are renewable drill bushings and liner bushings. The drill bushing locates and supports the cutting tool. The liner bushing locates and supports the drill bushing. Both the drill bushings and liner bushings for renewable arrangements are available in several styles.

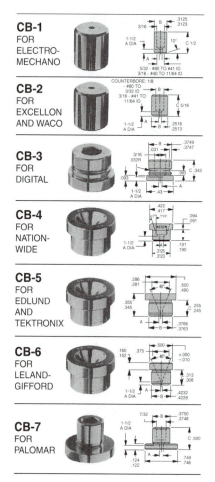

**Figure 7-14.** *Circuit-board bushing variations.*

*Slip/Fixed Renewable.* Slip/fixed renewable bushings, types SF and SFC (carbide), are the most-common form of renewable bushings, Figure 7-15. This renewable bushing is the replacement for the older

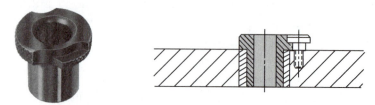

**Figure 7-15.** *Slip/fixed renewable bushings are replaceable drill bushings used in high-volume production.*

and obsolete slip renewable, type S, and fixed renewable, type F, bushings. Slip/fixed renewable bushings combine both the slip and fixed locking arrangements in the same bushing, Figure 7-16.

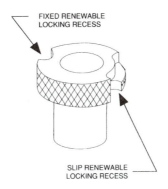

**Figure 7-16.** *The slip/fixed renewable bushings combine both slip and fixed locking arrangements on opposite sides of the bushing head.*

Slip/fixed renewable bushings are typically employed in long production runs where bushing changes are needed. These bushings can be installed in either a fixed-renewable or slip-renewable configuration by simply rotating the bushing, Figure 7-17.

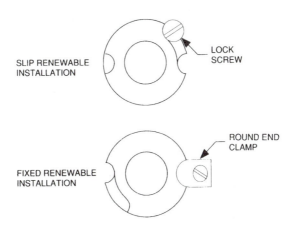

**Figure 7-17.** *The slip/fixed renewable bushing can be installed as either fixed-renewable or slip-renewable by simply rotating the bushing.*

The fixed-renewable installation is intended for single-step applications such as drilling or reaming. These bushings are changed only when the bushings wear. Fixed-renewable bushings are held in place with a lockscrew or round clamp. The clamps hold the bushing in place and prevent any movement during the machining cycle. When the bushing must be replaced, the clamp is removed and the bushing changed. The clamp is then reinstalled to securely hold the bushing.

Slip-renewable installations are convenient for applications when multiple operations are performed in the same hole. One example is drilling and reaming the same hole. The first slip-renewable bushing is installed, and the hole is drilled. The drilling bushing is removed. Then the reaming bushing is installed, and the hole reamed to size.

The slip-renewable side allows rapid changeover. The bushing is rotated clockwise to lock it in place and rotated counterclockwise for removal, Figure 7-18. A cutout at the end of the recess makes the bushing easy to remove and replace. Although slip/fixed renewable bush-

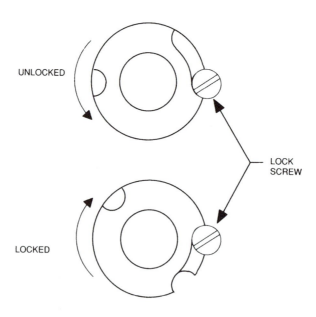

**Figure 7-18.** *The slip-renewable bushing is rotated clockwise to lock it in place, and rotated counterclockwise for removal.*

ings are normally installed in a liner bushing, they can also be installed directly in the jig plate. Figure 7-19 shows the recommended hole sizes for installing slip/fixed renewable bushings without a liner bushing.

| RECOMMENDED HOLE SIZES FOR SLIP/FIXED RENEWABLE BUSHINGS | | |
| --- | --- | --- |
| NOMINAL BUSHING O.D. | ACTUAL BUSHING O.D. | RECOMMENDED HOLE SIZE |
| 3/16 | .1875-.1873 | .1880-.1883 |
| 1/4 | .2500-.2498 | .2507-.2510 |
| 5/16 | .3125-.3123 | .3132-.3135 |
| 3/8 | .3750-.3748 | .3757-.3760 |
| 7/16 | .4375-.4373 | .4382-.4385 |
| 1/2 | .5000-.4998 | .5007-.5010 |
| 9/16 | .5625-.5623 | .5632-.5635 |
| 5/8 | .6250-.6248 | .6257-.6260 |
| 3/4 | .7500-.7498 | .7507-.7510 |
| 7/8 | .8750-.8748 | .8757-.8760 |
| 1 | 1.0000-.9998 | 1.0007-1.0010 |
| 1-3/8 | 1.3750-1.3747 | 1.3757-1.3760 |
| 1-3/4 | 1.7500-1.7497 | 1.7507-1.7510 |
| 2-1/4 | 2.2500-2.2496 | 2.2507-2.2510 |

**Figure 7-19.** *Recommended hole sizes for installing slip/fixed renewable bushings without a liner bushing.*

*Liner.* Liner bushings, type L, Figure 7-20, resemble press-fit bushings, but are larger. Liner bushings are used with the renewable-style bushings to provide a hardened, wear-resistant hole in an otherwise-soft jig plate. The close sliding fit between the renewable bushing and the liner bushing permits the bushing to be changed repeatedly over long production runs with no loss of positional accuracy. The headless design of liner bushings allows them to be mounted close together and flush with the top of the jig plate. As with the press-fit-type bushing, however, these bushings offer less resistance to heavy axial loads.

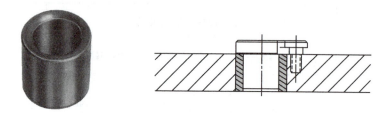

**Figure 7-20.** *Liners are permanent bushings used to hold and locate renewable drill bushings.*

*Head Liner.* Head-liner bushings, type HL, Figure 7-21, are similar to the liner bushing in design and application, but they are made with a head. Head-liner bushings, like head-press-fit bushings, are designed for applications where heavy axial loads might push a press-fit bushing through the mounting hole. These bushings can be mounted with the head exposed or counterbored, as shown. When the jig plate is counterbored for mounting, only the body diameter of the bushing provides

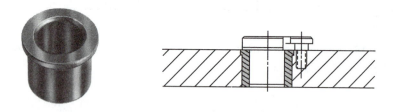

**Figure 7-21.** *Head liners have a head to resist heavy axial loads.*

the location and only this diameter needs to be reamed. The counterbored area provides clearance for the head and should not be a precision fit. Figure 7-22 shows standard head diameters. Note: the length of a head-liner bushing is measured from the top to bottom of the bushing and includes the height of the head.

| BODY DIAMETER | HEAD DIAMETER | HEAD THICKNESS |
|:---:|:---:|:---:|
| 1/2 | 5/8 | 3/32 |
| 3/4 | 7/8 | 3/32 |
| 1 | 1-1/8 | 1/8 |
| 1-3/8 | 1-1/2 | 1/8 |
| 1-3/4 | 1-7/8 | 3/16 |
| 2-1/4 | 2-3/8 | 3/16 |
| 2-3/4 | 2-7/8 | 3/16 |

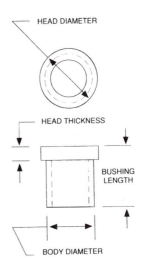

**Figure 7-22.** *Dimensions of head-type liner bushings.*

*Lockscrews and Clamps.* Renewable bushings are typically held in the jig plate with either a lockscrew or a clamp. The lockscrew or clamp both radially locates the bushing in the liner and holds the bushing in place. The lockscrew, Figure 7-23a, is the most-common form of locking device. These screws generally mount the bushings on either their slip-renewable or fixed-renewable sides. The screws are made with a shoulder under the head as shown. For mounting bushings on their slip-renewable side, the shoulder provides the necessary clearance needed to rotate the bushing for installation and removal. When mounted on the fixed-renewable side, the underside of the head securely holds the bushing in place.

The lockscrew locating jig, Figure 7-23b, locates these lockscrews with respect to the renewable bushing. As demonstrated in Figure 7-23c, the lockscrew locating jig is positioned against the bushing and struck with a hammer to mark the location of the lockscrew.

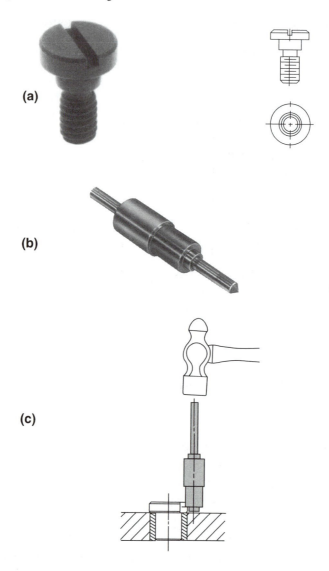

**(a)**

**(b)**

**(c)**

**Figure 7-23.** *Lockscrews are the most-common holding device for renewable bushings. The centerpunch shown is a multipurpose locating jig for most lockscrews and clamps.*

The round end clamp, Figure 7-24a, can also be used for mounting bushings on either their slip-renewable or fixed-renewable sides. This clamp comes in two heights for bushing installations with either a recessed or projected liner, Figure 7-24b.

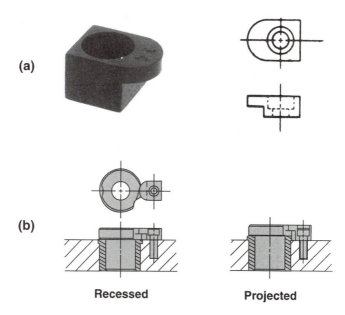

**(a)**

**(b)**

Recessed          Projected

**Figure 7-24.** *The round end clamp is a heavy-duty alternative to lockscrews.*

The round clamp is a bushing clamp designed specifically to hold bushings on their fixed-renewable side. As shown in Figure 7-25, round clamps are held in place with a socket-head cap screw. The lockscrew locating jig can also be used to install these clamps.

The flat clamp, shown in Figure 7-26a, is another form of bushing clamp. These clamps are used for the older styles of fixed-renewable bushings which have a flat-milled clamping area. Like the round end

**Figure 7-25.** *The round clamp is used to clamp bushings tightly on their fixed-renewable side.*

clamp, the flat clamp is made in two heights for bushing installations with either a recessed or projected liner, Figure 7-26b.

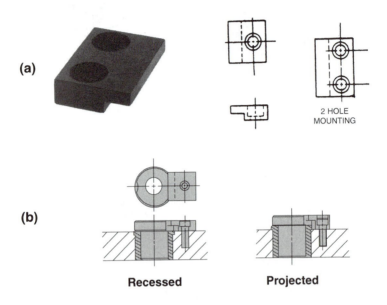

**(a)**

2 HOLE
MOUNTING

**(b)**

**Recessed**          **Projected**

**Figure 7-26.** *The flat clamp is a bushing clamp for the fixed-renewable bushings that have a flat-milled clamping area.*

*Locking Liner.* The locking liner bushing, type UL, Figure 7-27, is a unique bushing design for slip-renewable bushing installations. As shown, the bushing combines both liner and locking device in a single unit. The basic design of this bushing is similar to a head liner, but it has a special locking tab that eliminates the need for a lockscrew.

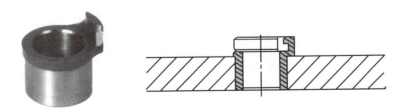

**Figure 7-27.** *Locking liner bushings include a special locking tab that eliminates the need for a lockscrew. These liners can only be used for slip-renewable applications, not fixed.*

Locking liner bushings are slightly more expensive than the head-liner-bushing/lockscrew unit they replace, but the reduced installation time more than offsets any additional cost. Note: these liners can only be used on the slip-renewable side.

*Diamond-Knurled Locking Liner.* The diamond-knurled locking liner bushing, type ULD, Figure 7-28, is a variation of the locking liner. These bushings are a form of liner bushing for cast-in-place applications. They combine both the liner and the locking device into a single unit, but the diamond-knurled locking liners have a diamond-pattern knurl on their circumference. The knurl offers high resistance to both rotational and axial forces. Like the other knurled bushings, the circumference of these liner bushings is not ground, so the internal diameter must be used to align the bushing for cast-in-place applications.

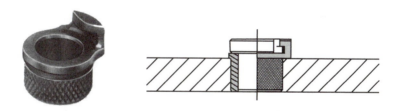

**Figure 7-28.** *Diamond-knurl locking liner bushings are cast in place or potted.*

*EZ-Cast Liner.* EZ-Cast liner bushings, type EZ, Figure 7-29, are another form of cast-in-place liner bushing. However, unlike the diamond-knurled locking liners, these bushings have an integral lockscrew and can be used to mount either side of slip/fixed renewable bushings.

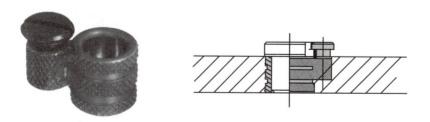

**Figure 7-29.** *EZ-Cast liner bushings are cast-in-place liners that can be used for either slip-or fixed-renewable applications.*

The headless design of these liner bushings permits them to be mounted flush with the top of the jig plate. The diamond-pattern knurl offers high resistance to both rotational and axial forces. Like other knurled bushings, the mounting surface of these bushings is not ground, so the internal diameter must be used to accurately align the bushing.

*Gun Drill.* Gun-drill bushings, Figure 7-30, are specialty bushings for deep-hole-drilling machines. The type of bushing is determined by the type of gun-drilling machine. Gun-drill bushings are similar to slip/fixed renewable bushings in both their appearance and application, but gun-drill bushings have the drill-bearing area at the head end of the bushing. Depending on the gun-drilling machine, these bushings are either one-piece or two-piece units.

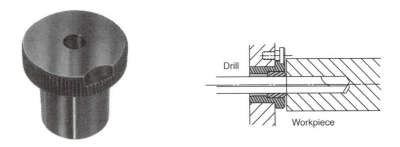

**Figure 7-30.** *Gun-drill bushings are specially designed for gun-drilling machines (deep-hole drilling).*

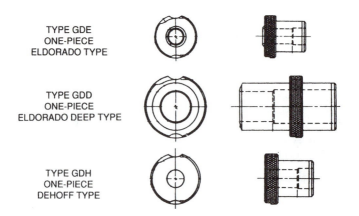

**Figure 7-31.** *Variations of one-piece gun-drill bushings.*

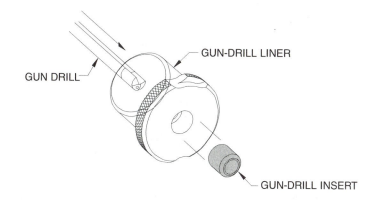

**Figure 7-32.** *The gun-drill liner and gun-drill insert bushings are used together.*

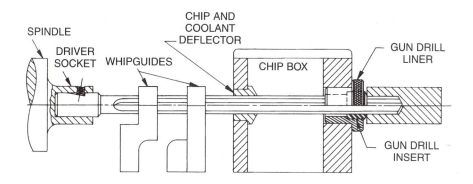

**Figure 7-33.** *Typical gun-drilling setup.*

The GD-type gun-drill bushing is a one-piece bushing. As shown in Figure 7-31, a third letter is added to the GD designation. This letter matches the bushing to a specific type of gun-drilling machine. The GDL (liner) and GDI (insert) bushings are two-piece units. As shown in Figure 7-32, these bushings are used together. Figure 7-33 shows the bushings in a typical gun-drilling setup.

## Air-Feed-Drill Bushings

Air-feed-drill bushings are special-purpose bushings designed for a variety of commercial self-feeding air-feed drills, tappers, and back-spotfacers. These drill bushings, called "shanks," are part of a complete system that includes the shanks, collars, and mounting devices. The shanks and collars are available either individually or as assembled units. When assembled, the unit is called an adaptor-tip assembly, Figure 7-34.

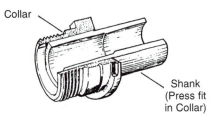

**Figure 7-34.** *Adaptor-tip assemblies are the drill bushings mounted on air-feed drills.*

The shanks, like standard bushings, are made with an internal diameter sized to fit the cutting-tool diameter, and an external diameter sized to fit the jig-mounted liner bushing. The collar is designed both to mount the tip assembly to the self-feeding drill motor, and to hold the complete unit in the jig-mounted liner bushing. As shown in Figure 7-35, the tip assembly is inverted in the liner and turned counterclockwise 30° to lock the units together. Figure 7-36 shows the various mounting options available for the jig-mounted liner bushings. The two primary

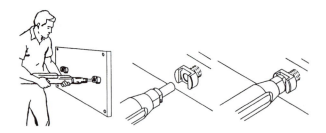

**Figure 7-35.** *Air-feed drills are mounted by inserting the adaptor tip into a locking liner.*

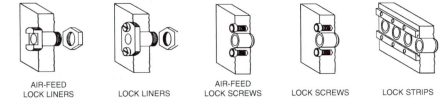

AIR-FEED
LOCK LINERS          LOCK LINERS          AIR-FEED
LOCK SCREWS          LOCK SCREWS          LOCK STRIPS

**Figure 7-36.** *Liner-mounting options for air-feed-drill bushings.*

forms of air-feed bushings are those for standard air-feed drills and those for coolant-type air-feed drills.

*Standard Bushings.* Standard air-feed bushings, Figure 7-37, have a tip assembly made from individual shank and collar units. The collar attaches the unit to the air-feed-drill nosepiece either directly or through an optional reducer. The shank can have a plain end or a contour-nose design. The contour-nose shank is a modified standard shank for appli-

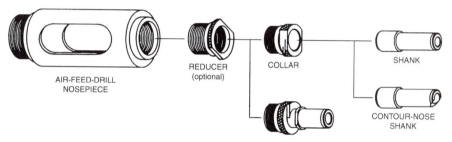

**Figure 7-37.** *The nose assembly of an air-feed drill.*

cations where curved or sloped surfaces are drilled. As shown in Figure 7-38, the contour-nose shank has a dual-angle angular relief, of 8° and 45°, on the end of the shank.

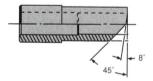

**Figure 7-38.** *The contour-nose shank is modified for drilling curved or sloped surfaces.*

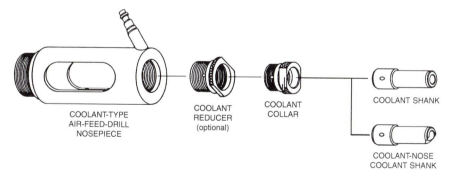

**Figure 7-39.** *Coolant-inducing bushings are similar to standard air-feed bushings, but have holes in each element to force coolant through the assembly to the cutting tool.*

*Coolant-Inducing Bushings.* Coolant-inducing bushings, Figure 7-39, are essentially the same as standard air-feed bushings, except for the addition of drilled passages in each of the elements designed to force coolant through the assembly to the cutting tool. The air-feed-drill nose-piece is also different, as shown, to fit the connector for the coolant hose.

## OPTIONAL FEATURES

In addition to the standard bushing variations, a range of optional features is also available for specific drilling situations. These optional features increase the versatility of the bushings and are useful in several ways.

### Chip Breakers

Chip-breaker bushings, type CH, Figure 7-40, are similar to the SF renewable-type bushings. These bushings, however, have a series of specially designed notches on the drill-exit end of the bushing. The notches break up chips created when tough or stringy materials are drilled. Breaking up the chips reduces friction and heat buildup. In addition, chip breakers also reduce wear on the drill-exit end of the bushing and minimize any chance of damage to either the bushing or workpiece. Chip breakers are also available on other bushing types, such as P and H bushings.

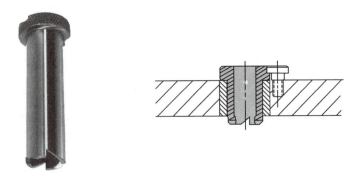

**Figure 7-40.** *Chip-breaker bushings have specially designed notches on the drill-exit end for breaking chips.*

## Directed Coolant Passages

Directed-coolant bushings, Figure 7-41, are also similar to the SF renewable bushings. Directed-coolant bushings, type DC, come with coolant passages machined into the bushings which direct the coolant flow to the cutting area. This design both cools the cutting tool and washes away the accumulated chips. Directed-coolant bushings can be

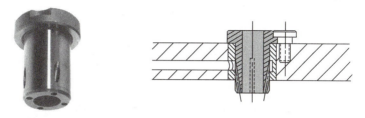

**Figure 7-41.** *Directed-coolant bushings have drilled passages to direct coolant to the cutting area.*

mounted in either DCL (liner) or DCHL (head liner) bushings, Figure 7-42. These special liner bushings have a unique design that directs the coolant flow from a drilled manifold passage, through the liner, to the holes in the bushing wall.

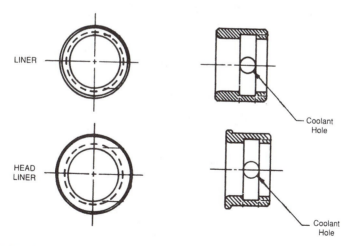

**Figure 7-42.** *Directed-coolant liners are used with directed-coolant bushings.*

## Oil Grooves

Oil-groove bushings, Figure 7-43, ensure adequate cutting-tool cooling and lubrication. This style bushing is well suited for operations such as drilling hardened steel, where a constant supply of cutting oil is required. Oil grooves are available in most bushing styles including the P, H, and SF renewable bushings. The grooves are specially designed passageways cut into internal diameter wall of the bushing.

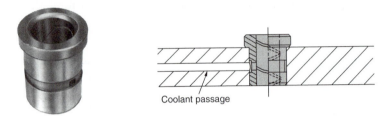

Coolant passage

**Figure 7-43.** *Oil-groove bushings have internal grooves that allow complete drill lubrication and cooling.*

Oil-groove bushings are made with either an oil hole, an oil hole and external groove, or without an oil hole, Figure 7-44. The bushings with oil holes direct the oil flow from a drilled manifold passage. Bush-

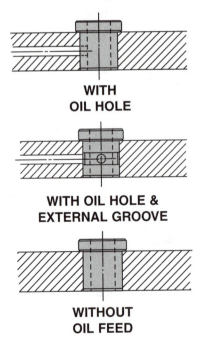

**WITH
OIL HOLE**

**WITH OIL HOLE &
EXTERNAL GROOVE**

**WITHOUT
OIL FEED**

**Figure 7-44.** *There are three general options for supplying fluid to the internal grooves.*

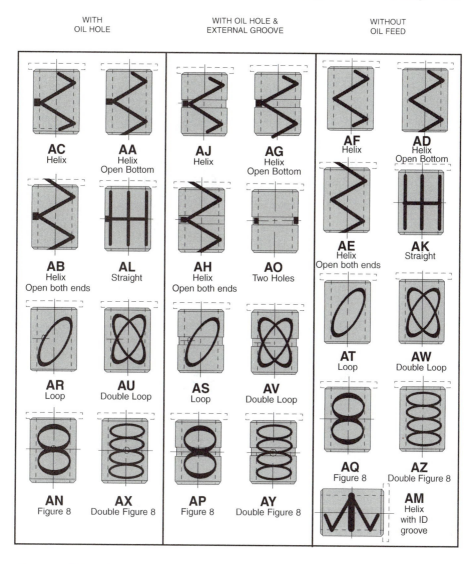

WITH
OIL HOLE

WITH OIL HOLE &
EXTERNAL GROOVE

WITHOUT
OIL FEED

**Figure 7-45.** *Oil-groove bushings are available in 25 different groove styles.*

ings without an oil-feed hole use gravity to feed oil to the cutting tool through the head end of the bushing. There are 25 different groove patterns to fit virtually any requirement, Figure 7-45. End wipers, Figure 7-46, are also available to keep out dirt and chips. These wipers are for oil-groove patterns that do not break out at the wiper end.

## OPTIONAL WIPERS

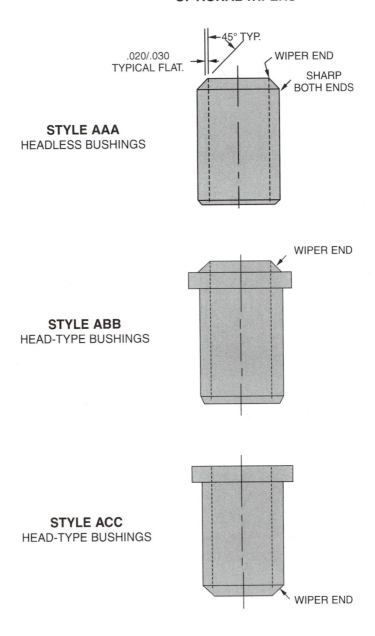

.020/.030
TYPICAL FLAT.

45° TYP.

WIPER END

SHARP
BOTH ENDS

**STYLE AAA**
HEADLESS BUSHINGS

WIPER END

**STYLE ABB**
HEAD-TYPE BUSHINGS

**STYLE ACC**
HEAD-TYPE BUSHINGS

WIPER END

**Figure 7-46.** *Oil-groove bushings are optionally available with wipers to guard against dirt and chips entering the inside.*

## Angled Exit Ends

Bushings should always be positioned so they offer maximum support for the cutting tool. Occasionally, with odd-shaped surfaces, this is not possible with standard bushings. So the exit end of the bushing must be modified or altered to match the specific shape of the workpiece surface. Bushings with angled exit ends, Figure 7-47, are available for such conditions. Altering the shape of the exit end of the bushing affords the best support. It also prevents any lateral movement, or wandering, of the cutting tool. Although almost-any-size cutting tool can move off the desired center position if improperly supported, this is especially true with smaller-diameter cutting tools.

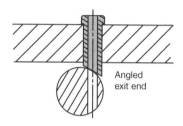

**Figure 7-47.** *Angled-exit-end bushings are sometimes required, for drilling curved or sloped surfaces.*

## Ground Flats

Ground flats are typically specified for bushings that must be positioned close to each other in the jig plate. Although almost any bushings can use ground flats, they are especially appropriate for headed bushings, as shown in Figure 7-48. The ground flats permit standard bushings to be positioned very close together with a simple alteration.

**Figure 7-48.** *Ground flats can be used for positioning bushings close together.*

## Thinwall Bushings

Thinwall bushings, as their name implies, are drill bushings made with a very-thin wall, Figure 7-49. These bushings are also used for applications where holes are close together. One note of caution: Since the wall thickness is quite thin, these bushings will tend to follow the shape of the mounting hole. For this reason, the geometry of the mounting hole is very important to the accuracy of the installation.

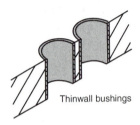

Thinwall bushings

**Figure 7-49.** *Thinwall bushings can also be used for positioning bushings close together.*

## Alternate Materials

Standard drill bushings are made of 1144 Stressproof steel. These are hardened to an RC 62-64 inside-diameter hardness. Other materials, such as 52100 steel, 300-series stainless steel, 400-series stainless steel, A2 tool steel, D2 tool steel, D3 tool steel, M2 tool steel, tungsten carbide, and bronze are also available for special situations.

## Special Bushing Sizes

In addition to the wide range of standard bushing sizes, virtually any combinations of inside diameter, outside diameter, length, head size, head style, or special tolerances are readily available as specials. These special bushings can be totally customized to fit any special machining situation or need.

## INSTALLATION

Drill bushings must be properly installed to do their job. The installation starts with a careful design process which matches the bushing type and size with the required operations. This process also involves selecting the correct jig-plate thickness and establishing the proper mounting clearance between the bushing and workpiece.

## Jig Plates

"Jig plate" is the term used to identify the parts of a jig which hold and support the drill bushings. The thickness of the jig plate is an important consideration in the installation of drill bushings. The jig-plate thickness is usually determined by the sizes of the bushings required.

As a general rule, bushings should be only long enough to properly support and guide the cutting tool. As shown in Figure 7-50, the thickness of the jig plate should generally be one to two times the tool diameter. This thickness provides adequate support for the cutting tool, yet keeps the jig plate as light as possible. When several different drill sizes are used, jig-plate thickness is normally determined by the largest tool diameter.

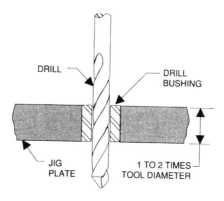

**Figure 7-50.** *Jig-plate thickness should be 1 to 2 times the tool diameter.*

## Chip Clearance

Chip clearance is another factor that must be carefully thought out before selecting and installing drill bushings. Chip clearance is the distance between the end of the bushing and the surface to be machined. As a rule, materials or operations that produce large stringy chips normally require greater clearance. Those producing small chips need less clearance.

In most cases, installations with little-or-no clearance will position the tool more accurately, but have a tendency to clog with chips. Also, when a bushing is positioned against the workpiece, the actual bearing area of the drill in the bushing is reduced by the length of the drill point. Too large a clearance, on the other hand, while less likely to clog, can also increase the chance of positional inaccuracy.

As shown in Figure 7-51, the recommended clearance for general-purpose drilling is 1/2 to 1-1/2 times the tool diameter. This clearance between the bushing and workpiece reduces chip interference. An operation such as reaming, which produces smaller chips and requires greater positional accuracy, should generally have a clearance of approximately 1/4 to 1/2 the tool diameter. In these situations, the smaller chips are less of a problem and permit a reduced clearance to ensure the necessary accuracy.

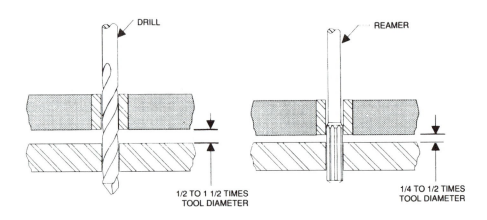

**Figure 7-51.** *The chip clearance between the bushing and the workpiece should be 1/2 to 1-1/2 times the tool diameter for drilling, and 1/4 to 1/2 times the tool diameter for reaming.*

When both drilling and reaming are performed in the same location with a renewable bushing, two different bushing clearances can be used. A bushing arrangement such as shown in Figure 7-52, slip/fixed renewable bushings with two different lengths, meets both clearance requirements. The shorter bushing is used for drilling; the longer bushing is used for reaming.

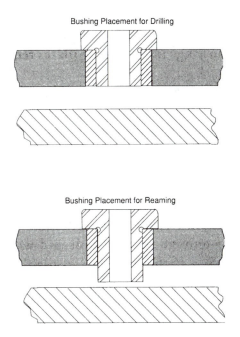

**Figure 7-52.** *When holes are first drilled and then reamed, different-length slip-renewable bushings can be used.*

## Installation-Hole Preparation

The first considerations in drill-bushing installation are the size and geometry of the mounting holes. All mounting holes must be perfectly round. For this reason, the holes should be jig bored or reamed to ensure the correct roundness. Ordinary twist drills should not be used for the finished mounting holes because they seldom cut a truly round or exact-size hole. The size of the hole is critical if the bushing is to perform properly.

The ideal interference, or press fit, is .0005" - .0008" for liners and headless press-fit bushings, and .0003" - .0005" for head-type press-fit bushings. A greater interference fit can cause problems by either distorting the bushing or deforming the jig plate. Less than the recommended interference fit can result in a loose-fitting bushing that spins when a load is applied.

Other factors to consider in determining the size of a mounting hole are:

- Head-type bushings require less interference to resist drilling thrust.
- Longer bushings in thicker jig plates require less interference.
- Thinwall bushings are more prone to distortion than normal bushings.
- Less-ductile jig-plate materials require less interference.

## Installation Procedure

For drill-bushing installation, an arbor press is preferred, Figure 7-53a. Using an arbor press permits a constant and even pressure. If an arbor press is unavailable and the bushing diameter is large enough, a draw-bolt and washers can also be used, Figure 7-53b. If the only tool available is a hammer, do not strike the bushing directly. This could fracture the bushing. Use a soft-metal punch to cushion the hammer blows.

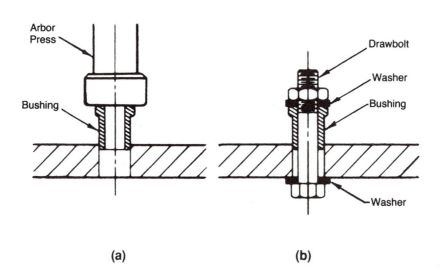

**(a)**                    **(b)**

**Figure 7-53.** *Two methods for properly installing press-fit bushings.*

To prevent damage to the bushing or mounting hole, coat both the inside surface of the mounting hole and the outside of the bushing with a lubricating compound. This makes the installation easier and reduces any scoring or galling between the mating surfaces. Always install the bushing with the ground lead entering the hole first. This lead helps align and position the bushing for installation.

## Unground Bushings for Oversize Holes

Each style of press-fit bushing is also available with an unground outside diameter. These are intended for customized applications in mounting holes that are worn or otherwise oversize. The specific amount of grinding stock left is determined by the size and style of bushing, but ranges from .005" - .020". The bushings are specified by adding the letter "U" in the standard bushing designation, e.g. H-20-16U-.1250. When grinding the bushings to a specific size, a grinding mandrel should be used to ensure the required concentricity of the inside and outside diameters.

# 8

# JIG & FIXTURE CONSTRUCTION

To get the greatest benefit from jigs and fixtures, a basic understanding of their construction is necessary. Jigs and fixtures are identified one of two ways: either by the machine with which they are used,
or by their basic construction. A jig, for instance, may be referred to as
a "drill jig." But if it is made from a flat plate, it may also be called a
"plate jig." Likewise, a mill fixture made from an angle plate may also
be called an "angle-plate fixture." The best place to begin a discussion
of jig and fixture construction is with the base element of all workholders, the tool body.

## TOOL BODIES

The tool body provides the mounting area for all the locators,
clamps, supports, and other devices that position and hold the workpiece. The specific design and construction of a tool body are normally determined by the workpiece, the operations to be performed, and
the production volume. Economy is also a key element in good design.

The three general categories of tool bodies today are cast, welded, and built up, Figure 8-1. Each type of construction can be used for
any workpiece, but one is often a better choice than the others. The first

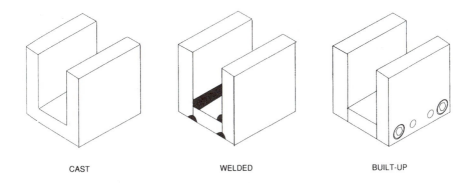

CAST           WELDED           BUILT-UP

**Figure 8-1.** *The three basic forms of tool-body construction are cast, welded, and built-up.*

step towards an economic design is to know and weigh the strengths and weaknesses of each.

Cast tool bodies are made in a variety of styles and types. The most-common casting materials for tool bodies include cast iron, cast aluminum, and cast magnesium. Cast materials occasionally found in specialized elements for tool bodies, rather than in complete tool bodies, are low-melting-point alloys and epoxy resins.

Cast tool bodies can have complex and detailed shapes. Such shapes require fewer secondary machining operations. Cast materials dampen vibration. They are most often found in relatively permanent workholders, workholders not subject to drastic changes. Cast tool bodies have three major drawbacks: (1) they are not easily modified for part changes; (2) their fabrication cost is high; (3) they require a lengthy lead time between design and finished tool body.

Welded tool bodies are also made from a wide variety of materials. The most-common welded tool bodies are made of steel and aluminum. Welded tool bodies are inexpensive to build, and they are usually easy to modify. They require minimal lead time. Welded tool bodies are also quite durable and rigid. They provide an excellent strength-to-weight ratio.

Heat distortion is the major problem with welded tool bodies. The secondary machining needed to remove distorted areas adds to the total cost of the workholder. Another problem with welded tool bodies is in the use of dissimilar materials. When a steel block, for example, is added to an aluminum tool body, it should usually be attached with threaded fasteners rather than welded to the body.

Built-up tool bodies are the most-common tool body today. These tool bodies are very easy to build, and usually require the least amount of lead time between design and finished tool. The built-up tool body is also easy to modify for changes in the part design. Like the welded body, built-up tools are durable and rigid, and have a good strength-to-weight ratio. Depending on the complexity of the design, the built-up tool body may be the least expensive to construct.

Built-up tool bodies are usually made of individual elements, assembled with screws and dowel pins. The built-up tool body is often used for precision machining operations, inspection tools, and some assembly tooling.

Preformed materials can often reduce the cost of machining tool bodies. These preformed materials include precision tooling plates, tooling blocks, cast sections, and angle brackets. Other materials include ground flat stock, drill rod, or drill blanks, and also structural sections such as steel angles, channels, or beams. The major advantage to using preformed and standard parts is the reduced labor cost in fabricating the workholder.

## Tooling Plates

Tooling plates are standard, commercially available base elements used to construct a variety of different workholders. Like other fixturing elements, these plates come in several variations to meet most fixturing requirements.

*Mill Fixture Bases.* Mill fixture bases, Figure 8-2, are commonly used for small-to-medium-size workholders. Although called "mill fixture" bases, these tooling plates can be used for many workholders, not just mill fixtures. Mill fixture bases are made in two general styles, standard and gang-milling. The standard mill fixture base is made in four different sizes ranging from 6" x 9" up to 12" x 18". The gang-milling fixture base comes in three sizes, from 6" x 9" up to 10" x 15".

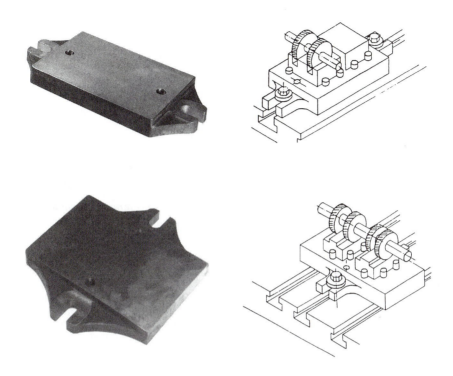

**Figure 8-2.** *Mill fixture bases provide an ideal start for small-to-medium-size milling fixtures.*

As shown, the major difference between these two fixture plates is in the location of the mounting lugs. The standard base has mounting lugs attached to the short side of the base. The gang-mill base has lugs mounted on the long side. Both styles of plates are made of ASTM Class 40 cast iron, normalized and ground flat and parallel to within .002". Both bases are also furnished with reamed mounting holes for Sure-Lock™ fixture keys to align the fixture bases to table T slots.

*Rectangular Tooling Plates.* Of all standard tooling plates, the rectangular ones, Figure 8-3, are the most popular. Their rectangular form works well for a wide variety of workholders. The plates come in a wide range of sizes, ranging from 12" x 16" to 24" x 32". Rectangular tooling plates are made of ASTM Class 40 gray cast iron, machined flat and parallel.

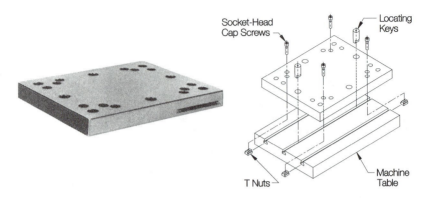

**Figure 8-3.** *Rectangular tooling plates are available in a full range of standard sizes, for vertical milling machines.*

Quick-change tooling plates, Figure 8-4, are a thinner variation of the rectangular tooling plate. They are made of 1045 steel and are precisely ground on both sides. These plates come in several sizes ranging from 5.50" x 7.50" to 9.50" x 17.50", and are designed to mount on standard tooling blocks.

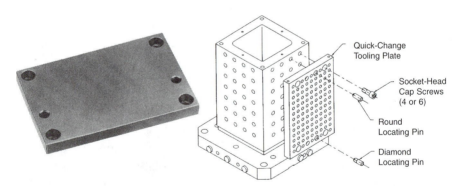

**Figure 8-4.** *Quick-change tooling plates are removable tooling plates that mount vertically on tooling blocks.*

*Round Tooling Plates.* Another tooling-plate variation is the round tooling plate, Figure 8-5. Round tooling plates work well on rotary or indexing tables. These tooling plates are available in 400mm, 500mm, and 600mm diameters. They are made of ASTM Class 40 gray cast iron, and have a series of mounting holes.

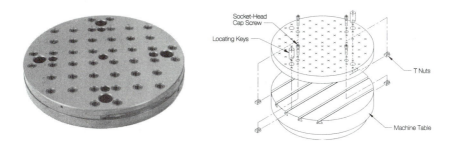

**Figure 8-5.** *Round tooling plates are ideal for rotary or indexing tables.*

*Square Pallet Tooling Plates.* Square pallet tooling plates, Figure 8-6, are another form of tooling plate. The square shape is ideal for palletized arrangements where a square tooling base is necessary. Square tooling plates come in five sizes to fit standard machining-center pallets 320mm, 400mm, 500mm, 630mm, and 800mm square. Plates are made of ASTM Class 40 gray cast iron.

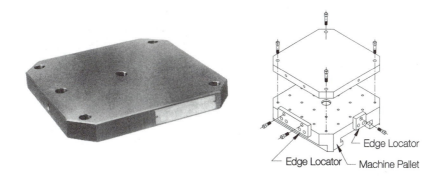

**Figure 8-6.** *Square pallet tooling plates are available for all horizontal machining centers with standard square pallets.*

*Rectangular Pallet Tooling Plates.* Similar to square pallet tooling plates, except made for rectangular machining-center pallets 320 x 400mm, 400 x 500mm, 500 x 630mm, and 630 x 800mm. Figure 8-7 shows this type, and how it can also be mounted on a square pallet by adding a spacer.

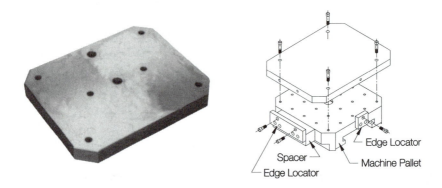

**Figure 8-7.** *Rectangular pallet tooling plates are for machining centers with rectangular pallets, or square pallets requiring a larger mounting surface.*

*Platform Tooling Plates.* Platform tooling plates are a variation of the square tooling plate. These plates are specifically designed for a mounting surface that must be elevated off the machine-tool table. As shown in Figure 8-8, the raised mounting surface permits easier access to the workpiece with horizontal machining centers. The added height provides the necessary clearance for the machine-tool spindle. The design also eliminates the dead space between the machine-tool table and the minimum operating height of the spindle. Platform tooling plates come in three sizes for 500mm, 630mm, and 800mm pallets. Platform tooling plates are made of ASTM Class 45 cast iron.

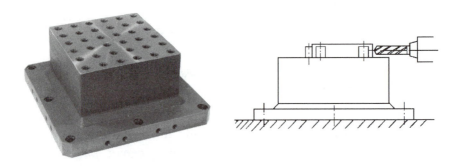

**Figure 8-8.** *Platform tooling plates provide a raised horizontal mounting surface for easier workpiece access on horizontal machining centers.*

*Angle Tooling Plates.* The angle tooling plates, Figure 8-9, are another useful tooling plate. These vertical plates allow mounting a large part approximately on the pallet's centerline. These plates are made to fit machining-center pallets 400mm, 500mm, 630mm, and 800mm square. Angle tooling plates are made from ASTM Class 45 cast iron.

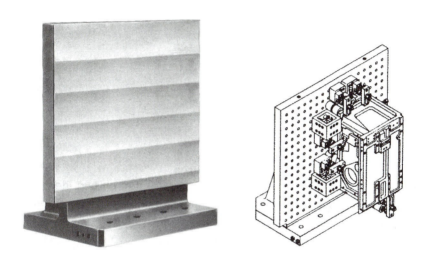

**Figure 8-9.** *Angle tooling plates provide a vertical mounting surface on horizontal machining centers, ideal for extremely large parts.*

## Tooling Blocks

Tooling blocks are often used on horizontal machining centers. The most-common tooling blocks are the two-sided and four-sided styles. These blocks work both for mounting workpieces directly, or for mounting other workholders. All working faces are accurately finish machined to tight tolerances, and qualified to the base. Dual mounting capability, Figure 8-10, allows both JIS mounting (locating from two reference edges) and DIN mounting (locating from center and radial holes).

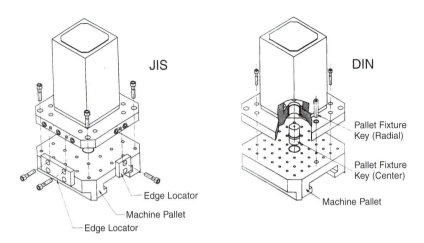

**Figure 8-10.** *Tooling blocks can be located either using two reference edges (JIS standard) or using center and radial holes (DIN standard).*

*Two-Sided Tooling Blocks.* The two-sided tooling block, Figure 8-11, is for mounting workpieces or workholders on two opposite sides. Two-sided tooling blocks work well for fixturing two large workpieces. These tooling blocks come in five different pallet sizes: 320mm, 400mm, 500mm, 630mm, and 800mm.

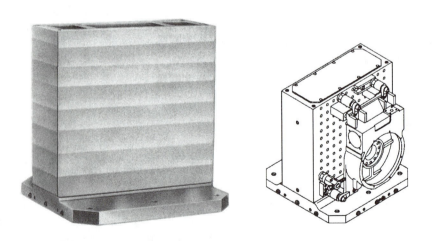

**Figure 8-11.** *Two-sided tooling blocks have two identical wide mounting surfaces for fixturing large parts.*

*Four-Sided Tooling Blocks.* The four-sided tooling block, Figure 8-12, mounts workpieces or workholders on four identical sides. Four-sided tooling blocks, with their four working surfaces, are typically chosen to maximize production. These tooling blocks are available in five different pallet sizes: 320mm, 400mm, 500mm, 630mm, and 800mm.

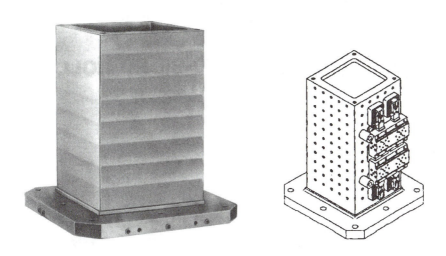

**Figure 8-12.** *Four-sided tooling blocks have four identical mounting surfaces for fixturing medium-size parts.*

## Precision Cast Sections

Precision cast sections come in a variety of shapes and sizes. Cast sections are available in standard lengths of 25.00" with squareness and parallelism within .005"/foot on all working surfaces. The sections can also be ordered cut to any specified length. The two common cast-section materials are cast iron and cast aluminum. The cast iron sections are made of ASTM Class 40 cast iron with a tensile strength of 40,000 to 45,000 psi. The aluminum sections are 319 aluminum with a tensile strength of 30,000 psi. Cast elements are used mainly for major structural elements of jigs and fixtures rather than as accessory items. Depending on the workholder design, it is possible to build a complete workholder by simply combining different sections.

*T Sections.* T-shaped cast sections, as shown in Figure 8-13, are made in two different styles: equal T sections and offset T sections. Equal and offset T sections come in 25.00" lengths. As shown in Figure 8-14, both the equal and offset T sections have basically a square cross-sectional profile where width and height are the same. The major difference between the two styles, as shown, is the position of the vertical member in relation to the horizontal portion. The vertical member of the equal T section is positioned in the middle of the bottom portion. It is moved to one side on the offset T section. Both styles are available in five different sizes ranging from 3.00" x 3.00" to 8.00" x 8.00". The web thickness of these sections is proportional to the overall size, ranging from .63" to 1.25". Figure 8-15 shows an application where either style T section can be used.

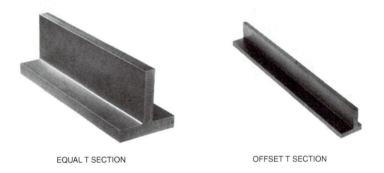

EQUAL T SECTION          OFFSET T SECTION

**Figure 8-13.** *T sections are made in two styles: equal T sections and offset T sections.*

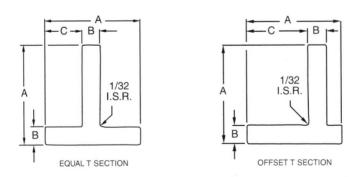

EQUAL T SECTION          OFFSET T SECTION

**Figure 8-14.** *The vertical member is positioned in the middle of an equal T section, while it is moved to one side on an offset T section.*

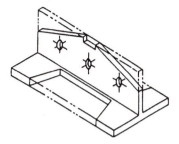

**Figure 8-15.** *An application where either T section may be used.*

*L Sections.* The L-shaped cast section, Figure 8-16, has a right-angle shape and is often used for applications when the bottom portion of a T section might get in the way. As shown, both the vertical and horizontal sides are the same. L sections come in five different sizes ranging from 3.00" x 3.00" to 8.00" x 8.00". The web thickness is proportional to the overall size. Figure 8-17 shows a fixturing application with the L section.

**Figure 8-16.** *The L section.*

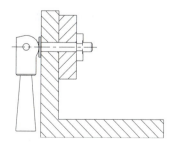

**Figure 8-17.** *A typical fixturing application using an L section.*

*U Sections.* U-shaped cast sections, Figure 8-18, are widely used for channel-type workholders. These sections have a square cross-sectional profile with identical height and width dimensions, as shown. U sections are available in seven sizes ranging from 1.75" square to 8" square. The web thickness of these sections is, once again, proportional to the overall size. The smaller U sections are made in 19.00" lengths. The larger sizes are available in full 25.00" lengths. Figure 8-19 shows two workholders constructed from this type of cast section.

**Figure 8-18.** *The U section.*

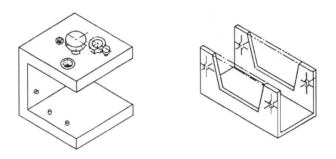

**Figure 8-19.** *Examples of workholders constructed from U sections.*

*V Sections.* V-shaped cast sections, Figure 8-20, are useful when a V-shaped element is needed for either locating or clamping. Thin portions of this material are often used as V pads. Longer lengths are frequently used as V blocks, Figure 8-21. V sections have a rectangular cross section with the width greater than the height. The V-shaped groove is machined to 90° ± 10'. V sections come in three sizes, ranging from 1.00" x 2.00" to 2.50" x 4.00" in standard 18.00" lengths.

**Figure 8-20.** *The V section.*

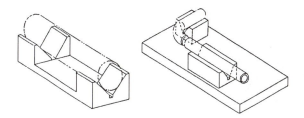

**Figure 8-21.** *V sections are often used as V blocks for locating cylindrical parts.*

*Square Sections.* Square cast sections, Figure 8-22, are typically used as major structural elements. Applications include riser elements, supports, or four-sided tooling blocks, as shown in Figure 8-23. Square sections are made in four standard sizes from 3.00" square to 8.00" square. All external surfaces except the ends are precisely machined.

**Figure 8-22.** *The square cast section.*

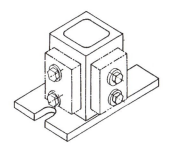

**Figure 8-23.** *A fixturing application with a square cast section used as a tooling block.*

*Rectangular Sections.* Rectangular cast sections, Figure 8-24, like square sections, are often used as structural elements in workholders. These sections work well for base elements, riser blocks, or similar features. Rectangular sections come in three standard sizes from 4.00" x 6.00" to 8.00" x 10.00". Here, too, all external surfaces except the ends are precisely machined.

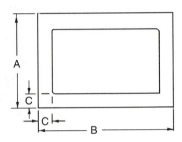

**Figure 8-24.** *The rectangular cast section.*

*H Sections.* H-shaped cast sections, Figure 8-25, are a unique design well suited for either complete tool bodies or structural elements. These sections are basically square and come in five sizes ranging from 3.00" x 3.00" to 8.00" x 8.00". All H sections are made in 25.00" length. Figure 8-26 shows an application with the H section as a tool body.

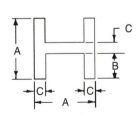

**Figure 8-25.** *The H section.*

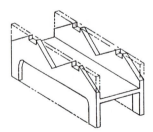

**Figure 8-26.** *A typical application with an H section as a tool body.*

*Flat Sections.* Flat cast sections, Figure 8-27, are the simplest and most-basic type of cast section. These sections are used where a cast iron material is preferred over steel flat stock. Flat sections are available in five width and thickness combinations. The sizes range from .63" x 3.00" to 1.25" x 8.00". Flat sections work well as base elements for smaller jigs or fixtures, or as structural elements for larger workholders.

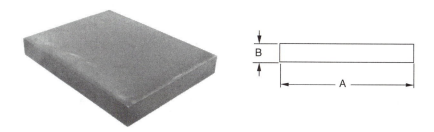

**Figure 8-27.** *The flat cast section.*

## Precision Angle Brackets

Angle brackets are often used when a right-angle alignment or reference is required. Although angle brackets are commonly thought of as 90° elements, there are also adjustable-angle styles of angle brackets and plates.

*Plain Angle Brackets.* The plain angle bracket, Figure 8-28, is available with or without locating holes. Angle brackets are often used when

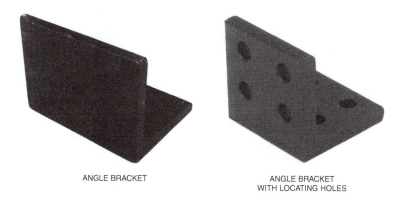

ANGLE BRACKET

ANGLE BRACKET
WITH LOCATING HOLES

**Figure 8-28.** *Angle brackets are machined flat and parallel to close tolerances. They are also available with precision locating holes for 3-axis accuracy.*

a fixed 90° angle is required. The right angle of these plates is closely controlled and is accurate to 90° ± .08°. These brackets are made in ASTM
A36 steel or 6061-T6 aluminum. Angle brackets come in ten different
sizes ranging from 2.00" x 2.50" to 6.00" x 6.00" with both equal and
unequal leg lengths. The web thickness of these sections is also proportional to the overall size, ranging from .22" to .44".

*Gusseted Angle Brackets.* The gusseted angle bracket, Figure 8-29,
is a variation of the standard angle bracket. This angle bracket is made
with a gusset between the horizontal and vertical legs. The gusset stiff-
ens the angle bracket and reduces any distortion when heavy loads are
applied. These angle brackets also have a right angle accurate to 90° ± .08°.
These brackets are made in ASTM A36 steel or 6061-T6 aluminum. Gusseted angle brackets are available in ten different sizes ranging from
2.00" x 3.00" to 6.00" x 6.00".

**Figure 8-29.** *Gusseted angle brackets have a gusset for added rigidity and strength.*

*Adjustable Angle Brackets.* Adjustable angle brackets, Figure 8-30,
are another variation of the plain angle bracket. These brackets are made
with a close-tolerance hinge between the horizontal and vertical legs.
The hinge permits the brackets to pivot so they may be set at any
desired angle. The most-basic adjustable angle bracket is the plain type,
shown at (a). This type has a bolt and nut arrangement for the hinge.
The gusseted adjustable angle bracket, shown at (b), also has a bolt and
nut hinge, but it also has a gusset mount on both legs. The mount permits the two legs to be connected with a gusset that is either bolted or

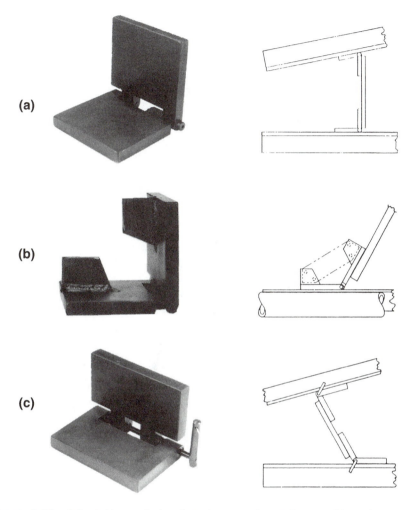

**(a)**

**(b)**

**(c)**

**Figure 8-30.** *Adjustable angle brackets have a close-tolerance hinge for accurate location.*

welded to the mounts. For applications where the angle bracket must be disassembled, the removable-pin-type adjustable angle can be used. This angle bracket, shown at (c), uses an L pin to attach the horizontal and vertical legs. These brackets are made of 1018 steel. The plain and gusseted adjustable angle brackets come in three different sizes ranging from 3.00" x 3.00" to 4.00" x 4.00", with equal or unequal leg lengths. The removable pin type is made in two sizes, 4.00" x 4.00" and 6.00" x 6.00".

$$F = \frac{W}{N \sin A}$$

If A = 70°, $F = \frac{4000}{4 \sin 70°} = 1064$ lbs

If A = 15°, $F = \frac{4000}{4 \sin 15°} = 3864$ lbs

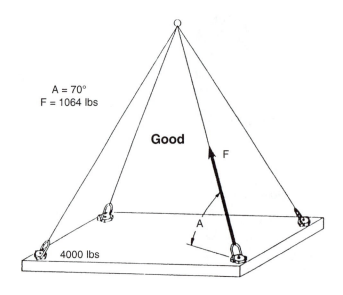

A = 70°
F = 1064 lbs

**Good**

F

A

4000 lbs

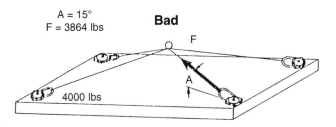

A = 15°
F = 3864 lbs    **Bad**

F

A

4000 lbs

**Figure 8-31.** *The load on a hoist ring is not simply total weight divided by the number of hoist rings. Shallow lift angles can cause very-large resultant forces.*

## Hoist Rings

Hoist rings should, for safety reasons, be added to any tool weighing over 30 pounds. The following are design considerations when selecting hoist rings.

*Hoist-Ring Safety Precautions.* Simply following a few basic safety precautions makes working with hoist rings both safer and more efficient.

1. The load on each hoist ring is not simply total weight divided by the number of hoist rings. The resultant forces can be significantly greater at shallow lift angles and with unevenly distributed loads. In the example shown in Figure 8-31:

$$F = \text{Force on each hoist ring}$$
$$W = \text{Total weight} = 4000 \text{ lbs}$$
$$N = \text{Number of hoist rings} = 4$$
$$A = \text{Lifting angle}$$

2. Despite the 5:1 safety factor on hoist rings, never exceed the rated load capacity. This safety margin is needed in case of misuse, which could drastically lower load capacity.

3. Tensile strength of fixture-plate material should be above 80,000 psi to achieve full load rating. For weaker material, consider through-hole mounting with a nut and washer on the other side.

4. Do not allow hoist rings to bind. Use a spreader bar, Figure 8-32, if necessary, to avoid binding.

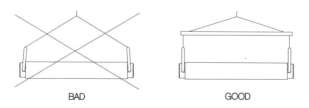

**Figure 8-32.** *Use a spreader bar to avoid binding hoist rings.*

5. Do not use spacers between the hoist ring and the mounting surface.

6. The mounting surface must be flat and smooth for full contact under the hoist ring. Tapped mounting holes must be perpendicular to the mounting surface.

7.  Tighten mounting screws to the torque recommended. Because screws can loosen in extended service, periodically check torque. For hoist rings not furnished with screws, mount only with high-quality socket-head cap screws.

8.  Never lift with a hook or other device that could deform the lifting ring. Use only cable designed for lifting.

9.  Do not apply shock loads. Always lift gradually. Repeat magnaflux testing if shock loading ever occurs.

10. After installation, check that rings rotate and pivot freely in all directions.

*Standard Hoist Rings.* Standard hoist rings, Figure 8-33, have a low profile and are attached directly to the mounting surface with socket-head cap screws. This is the moist-economical type of hoist ring. The solid forged lifting ring pivots 180° but does not rotate. These hoist rings are available for loads to 20,000 lbs.

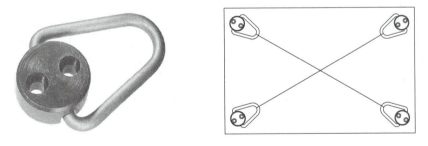

**Figure 8-33.** *Standard hoist rings have a forged ring that pivots 180°.*

**Figure 8-34.** *Swivel hoist rings pivot 180° and rotate 360° simultaneously to allow lifting from any direction.*

*Swivel Hoist Rings.* Swivel hoist rings, Figure 8-34, are a form of hoist ring with a 180° pivot and 360° rotation. These hoist rings, available in the two variations shown, are mounted with a single screw. As shown in Figure 8-35, these hoist rings are usually always preferred over conventional eye bolts when side loads are expected. The pivot-and-swivel combination permits the hoist ring to accommodate lifting angles that can cause a standard eye bolt to break. These rings are available for loads up to 10,000 lbs. They are available in a wide variety of sizes with either black oxide or cadmium plate finish.

A useful hoist-ring accessory is the hoist-ring clip, shown in Figure 8-36. These clips keep the swivel hoist rings stationary and out of the way when they are not being used for lifting.

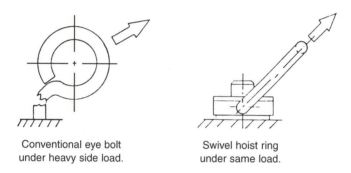

Conventional eye bolt
under heavy side load.

Swivel hoist ring
under same load.

**Figure 8-35.** *Hoist rings should be used in place of eye bolts for all heavy-lifting applications.*

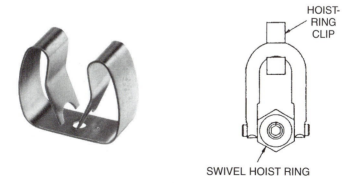

HOIST-
RING
CLIP

SWIVEL HOIST RING

**Figure 8-36.** *Hoist-ring clips keep swivel hoist rings stationary while not in use for lifting.*

*Lifting Pins.* Lifting pins, Figure 8-37, are a modified form of hoist ring. Sizes are available for loads up to 3400 lbs. These pins have a positive-locking four-ball arrangement to hold them in place during lifting. A release button at the opposite end of the pin allows quick installation and removal. Figure 8-38 shows two typical applications.

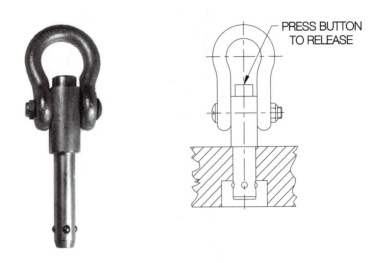

**Figure 8-37.** *Lifting pins are medium-duty hoist rings that can be completely removed.*

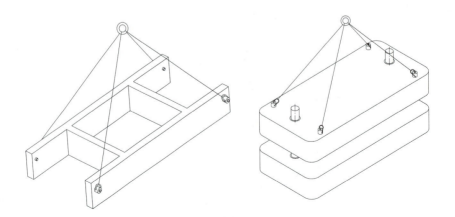

**Figure 8-38.** *Typical applications for lifting pins.*

## Threaded Inserts

Threaded inserts are widely used for construction and repair of workholders. The most-common use for the inserts is to reinforce threads in new workholders or to repair threads in existing workholders. Threaded inserts provide a way to strengthen threaded holes in new workholders where repeated use may cause excessive wear, such as with aluminum or other soft fixture plates. With existing jigs and fixtures, threaded inserts are a quick way to fix stripped, damaged, or worn threads. Two primary forms of threaded inserts are Keenserts and self-tapping threaded inserts.

Keenserts, Figure 8-39, are threaded inserts for both repairing and reinforcing applications. The Keensert design uses a standard tap size

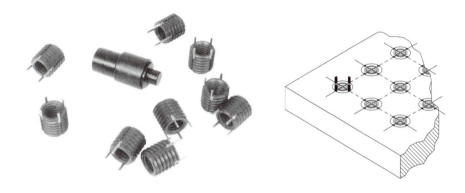

**Figure 8-39.** *Keensert threaded inserts are frequently used to reinforce threads in soft tooling plates.*

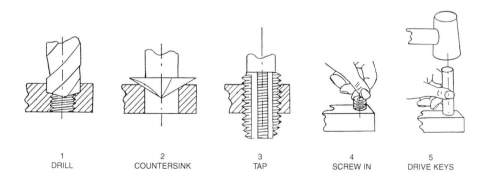

| 1 | 2 | 3 | 4 | 5 |
|---|---|---|---|---|
| DRILL | COUNTERSINK | TAP | SCREW IN | DRIVE KEYS |

**Figure 8-40.** *Installation steps for Keensert threaded inserts.*

for installing the insert. This feature eliminates the cost of special taps for threading the mounting holes. As shown, the inserts have unique locking keys that both securely hold the inserts in place and prevent rotational movement. The method of installing these inserts is shown in Figure 8-40.

1. Drill out the old thread, if repairing an existing thread, or drill a new hole to the specified tap drill diameter (slightly larger than the normal tap drill for that thread size).
2. Countersink the entry end of the hole to the specified diameter.
3. Tap the new threads to the correct size with a standard tap.
4. Screw in the insert until the body is slightly (.010" to .030") below the surface. The locking keys act as a depth stop.
5. Drive the keys down with several light hammer taps on the installation tool (or directly on the keys).

**Figure 8-41.** *Standard Keensert assortments contain many different insert sizes and a tool for each size.*

Keenserts are made in a variety of forms for almost any application. The inserts are available in a heavy-duty carbon-steel style in standard inch sizes from #10 through 1-1/2, and in metric sizes from M5 to M20. Stainless steel, heavy-duty, and thinwall inserts are also available in either plain or internal-thread-locking styles. A variety of standard Keensert assortments are also available for UNC, UNF and metric threads, Figure 8-41.

One other style of Keensert, ideal for repair work, is the solid Keensert, Figure 8-42. As shown, these stainless steel inserts are typically used for relocating tapped or drilled holes. They come with external-threaded diameters from 5/16" to 1-3/8".

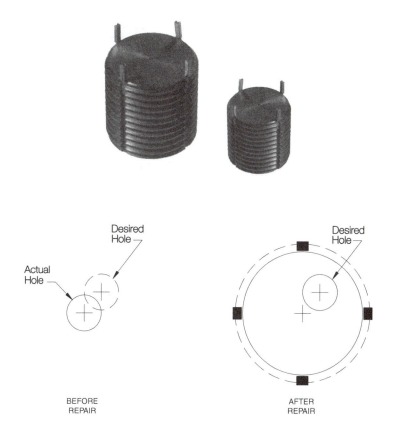

**Figure 8-42.** *Solid Keenserts are handy plugs for relocating misplaced holes.*

**Figure 8-43.** *Self-tapping threaded inserts can be used for aluminum or magnesium plates.*

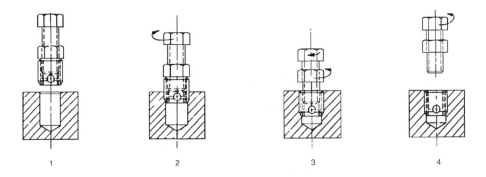

**Figure 8-44.** *Installation steps for self-tapping threaded inserts.*

The self-tapping threaded insert, Figure 8-43, is for permanent applications where threaded holes in either aluminum or magnesium require reinforcement. The hole is first drilled with the correct tap drill for the external thread on the insert. Then the inserts are installed following the steps shown in Figure 8-44.

1. Insert a screw in the insert and tighten a jam nut against the insert. The screw should not extend beyond the end of the insert.
2. Carefully align the insert with the hole. Turn the insert into the hole. The self-tapping feature of the insert cuts the necessary threads.
3. When the insert is installed to the desired depth, hold the screw with one wrench and break the jam nut loose with a second wrench.
4. To complete the installation, remove the screw and nut.

## Fixture Keys

An accurate relationship between the workholder and the machine tool must be established. Fixture keys not only establish location initially, they also help hold the fixture in place during machining.

The two basic styles of fixture keys are the slot-mounted and hole-mounted types. Slot-mounted fixture keys are made in two variations, the plain fixture key and the step fixture key, Figure 8-45. The plain fixture key, shown at (a), is the simplest and least-expensive of the slot-mounted keys. As shown, these keys are mounted in a slot cut to a depth equal to half the thickness of the key. The key is held in place with a socket-head cap screw.

The second style of slot-mounted fixture key is the step key, shown at (b). These fixture keys are a variation of the standard fixture key. This key's step design allows a fixture with one slot width to work on a machine table with a different slot width. Like the plain-style key, this key is held in place with a socket-head cap screw.

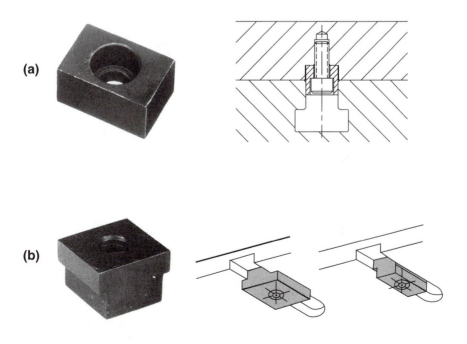

**(a)**

**(b)**

**Figure 8-45.** *Plain fixture keys and step fixture keys.*

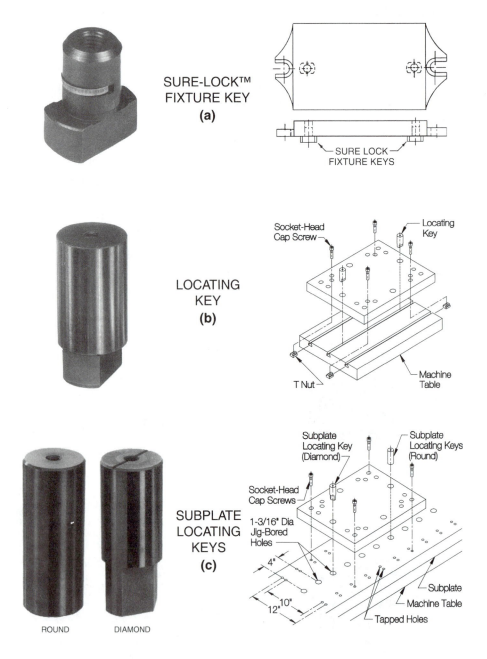

**Figure 8-46.** *Sure-Lock™ fixture keys are ideal for small and medium fixtures, while removable locating keys are best for large, heavy fixtures. Subplate locating keys allow mounting the same fixtures on a subplate with 3-axis location.*

Hole-mounted fixture keys are also made in several variations. The most popular are the Sure-Lock™ fixture key and the removable locating key, Figure 8-46. Hole-mounted keys eliminate the need to slot fixtures. The Sure-Lock™ fixture key, shown at (a), is the most popular of all fixture keys. Keys for any machine-table slot mount in a .6250" reamed hole for interchangeability. This design has a unique locking arrangement to precisely align and lock the fixture key in the hole. As shown, Sure-Lock™ fixture keys can be secured from either the top or bottom of the fixture. These keys are made in many sizes, for all standard USA table slots from 3/8" to 13/16", and metric table slots from 12mm to 22mm.

Locating keys, shown at (b), are the standard removable-type fixture key for large, heavy fixtures. These keys can be inserted from above after placing the fixture on the machine table, then removed again if desired after the fixture is fastened. This design keeps the fixture's bottom side free of obstructions. Locating keys all mount in a 1.1875" reamed hole. They are available in many sizes, for all standard USA table slots from 9/16" to 7/8", and metric table slots from 14mm to 22mm.

Subplate locating keys, shown at (c), are designed for mounting quick-change fixture plates on a subplate. A round key and a diamond (relieved) key are used together for precise location without binding. Two standard diameters, .6250" and 1.1875", match standard fixture-key holes. This allows mounting these same fixture plates either on a subplate (3-axis location) or directly on a machine table (2-axis location).

## JIG CONSTRUCTION

Jigs are made to meet the requirements and specifications of individual workpieces, resulting in an infinite number of different jigs. Even though every jig may be different, each can be grouped into one of only a few categories. The following is a description of the basic jigs and the applications where each is best.

### Template Jigs

Template jigs, Figure 8-47, are the simplest and least-expensive jigs. They are generally used for layout or light machining. They are designed for accuracy rather than speed. Template jigs do not have a self-contained clamping device. If a clamp is needed, a secondary clamp such as toggle pliers may hold the jig to the workpiece. When a template jig is used for drilling multiple holes, a pin placed in the first drilled hole can reference the jig.

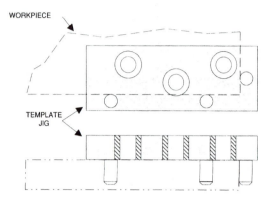

**Figure 8-47.** *Template jigs are simple plates usually held in place by hand.*

## Plate Jigs

Plate jigs are very similar in their basic construction to template jigs. As shown in Figure 8-48, plate jigs have a self-contained clamping device. Although many different clamps can be used, a screw clamp is the most-common type with these jigs.

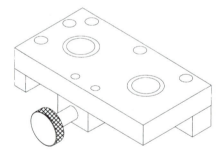

**Figure 8-48.** *Plate jigs are similar to template jigs, except they include a clamp.*

## Table Jigs

Table jigs are a variation of the basic template or plate jig. The table jig shown in Figure 8-49 is basically a template jig installed on legs. Table jigs are designed for applications where the surface to be machined also locates the workpiece. The location is transferred across the underside of the jig plate and down the legs to the machine table. When designing this type of jig, make sure the area to be drilled is inside the legs to prevent tipping. Although three legs will work, four are recommended with table jigs. Four legs always wobble if a chip is

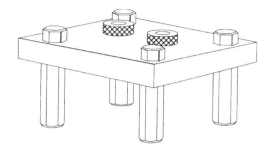

**Figure 8-49.** *Table jigs are basically template or plate jigs raised up on legs.*

under one leg; three legs do not. The wobble tells the operator to clear the chips from the locating surfaces.

## Sandwich Jigs

The sandwich jig, Figure 8-50, is another modified plate jig. These jigs are constructed with two jig plates. Depending on the design, sandwich jigs can machine a workpiece from either one side or both sides. This jig is designed for thin or delicate workpieces that require more support. When the machining is done from a single side, the second plate acts as a support and contains the necessary locating elements. When the machining is done from both sides, either plate can locate the workpiece. Locating pins and bushings are needed to maintain the proper relationship of the plates. The workpiece can be clamped with

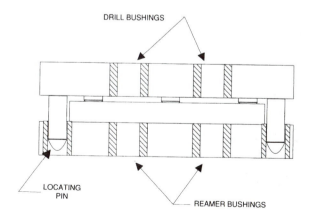

**Figure 8-50.** *Sandwich jigs can be used to drill from one side of the jig, then ream from the other side.*

either a side-mounted clamp screw, or as shown here, by clamping the two plates together against the workpiece.

## Box Jigs

Box jigs, or tumble jigs, are the most-detailed and most-complex jigs. They completely enclose the workpiece. Their construction provides for the workpiece to be completely machined on all sides without removal or repositioning. Box-jig bodies are available as off-the-shelf items, Figure 8-51. They can also be constructed from plate. Locators inside the box precisely position the workpiece. The top of the box jig is either removed, or it is hinged to load and unload the workpiece.

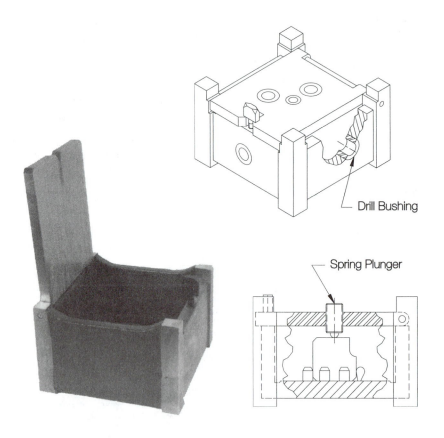

Drill Bushing

Spring Plunger

**Figure 8-51.** *Tumble box jigs allow drilling from any of six directions.*

## Leaf Jigs

Leaf jigs are simplified box jigs. They combine features of the box jig and sandwich jig. As shown in Figure 8-52, these jigs are available as ready-made units that require only locators. A leaf jig does not completely enclose the workpiece. When building leaf jigs, locators are usually mounted in the base to precisely position the workpiece. Drill bushings are installed in the leaf and/or the base. Additional elements can be added to the basic leaf jig to machine features on the sides of the workpiece.

## Pump Jigs

Pump jigs, Figure 8-53, are another ready-made tool body for drill jigs. These jigs are also referred to as Lift-N-C (Lift 'n See) jigs for the way that the top plate lifts up and opens wide for easy loading. These

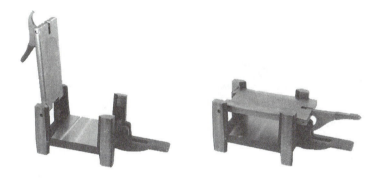

**Figure 8-52.** *Leaf jigs have a bushing-plate lid that swings open for easy loading.*

**Figure 8-53.** *Lift-N-C jigs get their name from the way that the bushing plate lifts up and opens wide for easy loading (lift 'n see).*

jigs act in much the same way as the leaf jig, but have a positive rack-and-pinion hinge arrangement. Pump jigs are normally used to machine a workpiece from only one side. When building a pump jig, the locating elements are mounted on the base to locate the workpiece. The drill bushings are installed in the top plate. When the jig is in the open position, the top plate is moved clear of the workpiece. This permits secondary operations, such as tapping, to be performed without removing the workpiece from the jig.

## Indexing Jigs

Indexing jigs are for applications where holes must be drilled in a pattern around a center axis. This is done either with an indexing ring holding bushings, or by indexing the part itself. With a separate indexing ring, Figure 8-54, a hand-retractable plunger is frequently used. A

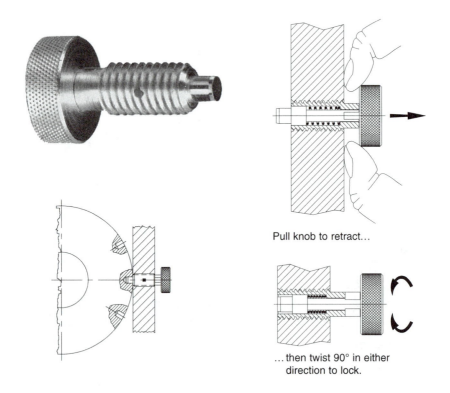

Pull knob to retract...

...then twist 90° in either direction to lock.

**Figure 8-54.** *A hand-retractable plunger can be used to positively index a ring holding drill bushings.*

ball-plunger can also be used for less-critical positioning. Here the workpiece itself serves as the indexing ring. With this design, the first hole is drilled and the part rotated to engage the ball-plunger, Figure 8-55. Once located, the part is reclamped and a second hole is drilled. The indexing is then repeated until all the holes in the part are drilled. The angular position of the ball plunger relative to the drill determines the indexing pattern. So, for four holes, the plunger is located at 90° from the drill, 60° for six holes, 45° for eight holes, and so on.

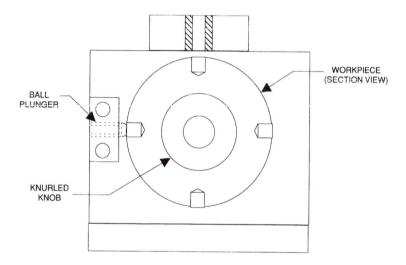

**Figure 8-55.** *A ball-plunger can index using holes drilled in the workpiece.*

## Multistation Jigs

Multistation jigs, Figure 8-56, are for repetitive simultaneous operations on several identical parts. In most cases, almost any jig may be used with a multistation arrangement. As shown, the unique feature of a multistation jig is the way the jigs are mounted and arranged with respect to the machining stations. In this example, the jig has four stations: #1 is the load/unload station; #2 is for drilling; #3 is the reaming station; and #4 is where the workpiece is counterbored. An indexing arrangement is also included with this jig to accurately position the jigs at each station.

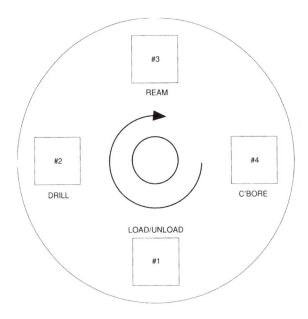

**Figure 8-56.** *Multistation jigs are used in a continuous, multi-step production process.*

## Trunnion Jigs

Trunnion jigs, Figure 8-57, are for large, heavy, or odd-shaped workpieces. This type of jig rotates the workpiece on precision bearing mounts called trunnions. Trunnions are made in two basic styles: standard and spherical.

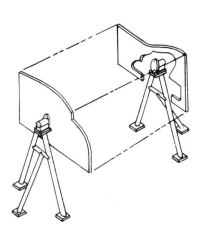

**Figure 8-57.** *Trunnion jigs allow easily turning a large part to work on all sides.*

Standard trunnions, Figure 8-58, are available in either locking or revolving styles. With most trunnion jigs, trunnions are typically used in pairs: the revolving trunnion provides a rotating pivot, and the locking trunnion both rotates and locks at any rotational angle. Locking trunnions are locked in place with a friction-cone arrangement engaged by turning a handle.

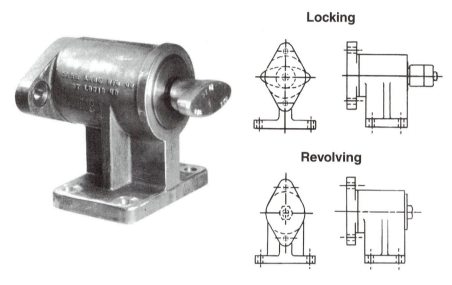

**Locking**

**Revolving**

**Figure 8-58.** *Trunnions are usually used in pairs, one locking and one revolving.*

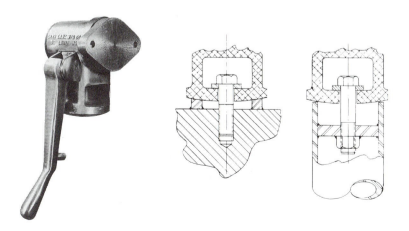

**Figure 8-59.** *Spherical trunnions are ideal for mounting on pipe frames.*

Spherical trunnions, Figure 8-59, are a locking trunnion for pipe-frame mounting. The spherical bottom aids in precisely aligning the trunnions. Spherical trunnions lock the jig in position by lowering the handle.

## FIXTURE CONSTRUCTION

Fixtures, like jigs, can be grouped into a few categories. These categories are most often based on the construction of the fixture. Another way to identify a fixture is by the machine it is used with.

### Plate Fixtures

The plate fixture is the most-basic and most-common fixture. The plate fixture is built with a mill fixture base, cast flat section, tooling plate, or similar plate material. All locators, supports, and clamps are mounted directly to the plate. As shown in Figure 8-60, a complete plate fixture can be built using only standard, off-the-shelf components.

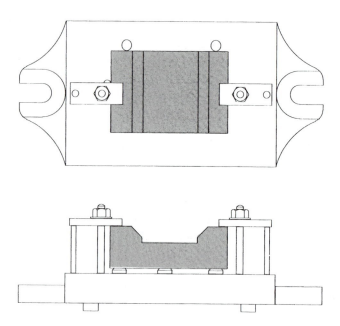

**Figure 8-60.** *Plate fixtures usually hold a workpiece parallel to the machine table.*

## Angle-Plate Fixtures

Angle-plate fixtures are a variation of the basic plate fixture. They are useful when the locating surface is at an angle to the machine table. The two main variations of angle-plate fixtures are the right-angle and modified-angle plate fixtures. Right-angle plate fixtures, Figure 8-61, are constructed at 90° to the base. The modified-angle plate fixtures have an angle other than 90°. The right-angle plate fixtures can be built with tooling blocks, T cast sections, L cast sections, angle brackets, or any comparable material. Adjustable angles or sine plates may be used to build the modified-angle plate fixtures.

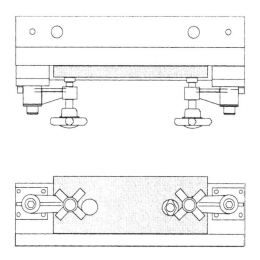

**Figure 8-61.** *Angle-plate fixtures usually hold a workpiece perpendicular to the machine table.*

## Welding Fixtures

All basic workholding principles should be applied to fixtures used for welding operations. The major differences between most welding fixtures and machining fixtures are the locational tolerances and clamping methods. With welding fixtures, weight is often a problem. Many times a fixture is made up of welded sections. The sections are usually positioned only in the areas where the parts to be joined contact the fixture. Rather than the precision locators found on most machining fixtures, small angle clips, blocks, or similar elements are used as locators.

The clamps for welding fixtures are often toggle clamps. These clamps offer the best combination of design flexibility, holding capacity, and operational speed. In addition, since most toggle clamps move completely clear of the work area when opened, loading and unloading operations are also simplified. Although the clamps may be attached with a screw, in many instances toggle clamps are welded directly to the fixture body.

A few important considerations to keep in mind when building a welding fixture are as follows:

- Always construct the fixture so that the parts to be welded can be loaded only in the correct orientation.
- Supports are needed underneath clamps to prevent distortion.
- Locators and supports must be positioned so any workpiece distortion caused by welding heat loosens, rather than tightens, the workpiece in the fixture.
- Only essential dimensions and relationships should be located and rigidly clamped.
- All areas to be welded must be easily accessible.
- When possible, welding should be done on a flat horizontal plane.
- To minimize warpage of the workpiece, provisions to dissipate excess heat must be included in the design.
- As many operations as possible should be performed before the workpiece is removed or repositioned.
- Large or heavy loads should be completely supported, and provisions for lifting the jig with a hoist must be included.

## Inspection Fixtures

Inspection and gaging fixtures are subject to different requirements than machining fixtures, whether inspection is done on a CMM (Coordinate Measuring Machine) or with manual gages. The following are key differences and unique aspects of inspection fixtures:

- The workpiece must be oriented to expose all features to be inspected.

- Machined surfaces are often not used for locating, to allow them to remain exposed.
- Bridge-type CMMs require different tooling than horizontal-arm machines.
- Since most CMMs can automatically reference a part to the machine, often only approximate orientation in the fixture is required.
- Quick-acting clamps, such as toggle clamps, are widely used in inspection fixtures due to the light clamping forces usually required.
- Inspection fixtures can be either permanent or modular.

Gaging fixtures are used to check or compare workpiece features against standards of known size. One often-used tool for gaging fixtures is the plug gage. Plug gages are available in two styles, the plain plug gage and the step plug gage.

*Plug Gages.* Plug gages are precision cylindrical gaging elements. They come in sizes from .011" to 1.000" in increments of .001" (.0005" increments are available as specials) in standard 2" lengths. These gages are available individually or in complete sets, Figure 8-62.

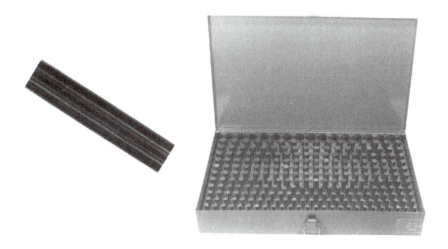

**Figure 8-62.** *Plug gages are available individually or in complete sets.*

Plug gages are used to check hole locations, distances between holes, hole sizes, and slot sizes, Figure 8-63. In most cases, plug gages are selected based on the upper and lower sizes of the feature to be checked. To check a hole, for example, the smaller plug gage is first inserted in the hole. Since the size of this gage matches the smallest size of the hole, the gage should enter the hole. If the gage is too large to enter the hole, then the hole is too small.  An attempt is then made to insert the larger gage. This gage matches the largest allowable size of the hole, so it should not be able to enter the hole. If the gage does enter the hole, then the hole is too large. Checking sizes with two plug gages is often called GO/NO-GO gaging. The plug gages for this method are often called GO and NO-GO gages. The GO/NO-GO gaging method is shown in Figure 8-64.

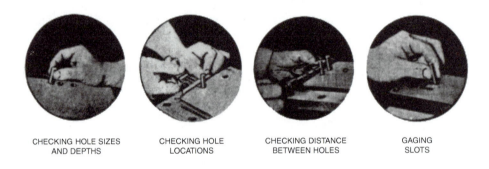

|  CHECKING HOLE SIZES<br>AND DEPTHS  |  CHECKING HOLE<br>LOCATIONS  |  CHECKING DISTANCE<br>BETWEEN HOLES  |  GAGING<br>SLOTS  |

**Figure 8-63.** *Applications of plug gages.*

*Step Plug Gages.* Step plug gages, Figure 8-65, are another plug gage used to check workpiece sizes and locations. As shown, the step plug gage is made with two diameters. This combines both the GO and NO-GO gaging elements into a single gage. The smaller diameter is the GO size and the larger diameter is the NO-GO size, Figure 8-66.

Although these gages can be used alone, a more-convenient way to use them is with a gage handle, Figure 8-67. Gage handles come in either single- or double-end styles, as shown. Single-end handles work with single plug gages or for step plug gages. The double-end handle

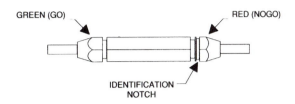

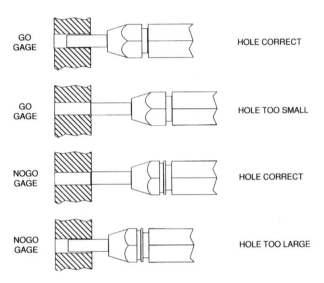

**Figure 8-64.** *The GO/NO-GO gaging method.*

**Figure 8-65.** *Step plug gages contain both the GO and NO-GO sizes.*

is designed for two gages for GO/NO-GO gaging. All handles have a small hole for a cable assembly, to allow attaching the gage permanently to an inspection fixture.

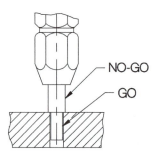

**Figure 8-66.** *Step plug gages allow checking GO and NO-GO tolerances simultaneously.*

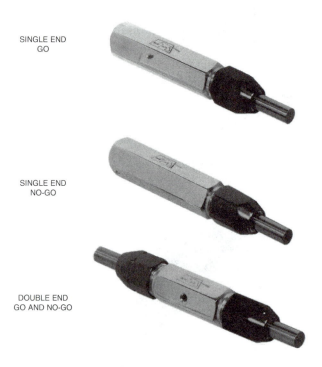

SINGLE END
GO

SINGLE END
NO-GO

DOUBLE END
GO AND NO-GO

**Figure 8-67.** *Plug gage handles have a drilled hole for attaching gages to an inspection fixture with a cable.*

# 9

# GENERAL-PURPOSE WORKHOLDERS

General-purpose workholders are the simplest and most-basic workholding devices. They are also the most-universal workholders. General-purpose workholders include chucks, collet chucks, collet vises, machine vises, and self-centering vises. They usually hold regular or symmetrically shaped workpieces: squares, rectangles, cylinders, hexagons, and similar part shapes. Although they lack the ability to hold specialized part shapes, they are still widely used. With modified clamping elements, many general-purpose workholders can be adapted for special part shapes.

## CHUCKS AND CHUCK JAWS

Chucks are commercial workholding devices for holding a variety of workpiece shapes, usually with three jaws, but also with two or four jaws. Although the standard chuck jaws furnished with most chucks offer some workholding options, designing a set of modified jaws offers almost unlimited options. A principal advantage of chucks for fixturing is the universal capability of these devices. Virtually any part within the size capacity of a chuck can be fixtured. Likewise, since only the jaws are modified for different workpieces, one chuck can fixture any number of different workpieces. Standard chucks can substantially reduce the design and construction cost of many workholders.

Chuck jaws for special chucking operations include two basic styles: hard jaws and soft jaws. Each of these styles comes in a wide variety of sizes and shapes.

## Hard Jaws

Single-step hard jaws are usually furnished with new chucks. These jaws are generally the most-universal type of chuck jaws. But even these jaws must occasionally be replaced. They are used in sets to match the number of jaws on the chuck. Figure 9-1 shows three styles of replacement hard jaws. These jaws are made to fit a wide variety of standard chucks and are available in three styles: two-step reversible, single-step reversible, and single-step non-reversible.

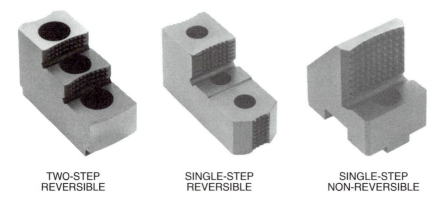

| TWO-STEP REVERSIBLE | SINGLE-STEP REVERSIBLE | SINGLE-STEP NON-REVERSIBLE |

**Figure 9-1.** *Hard jaws are general-purpose chuck jaws, with one or two ground steps on which to rest parts. Most are also reversible for through-feeding barstock.*

## Soft Jaws

Soft jaws, Figure 9-2, as their name implies, are usually made of carbon steel or aluminum. Like hard jaws, they are used in sets to match the number of jaws on the chuck. Soft jaws normally come in two heights, standard and extra high. The specific heights of each set depend on the type and size of chuck, and the jaw material. Soft jaws are partially machined to match specific master jaws for a wide variety of standard chucks.

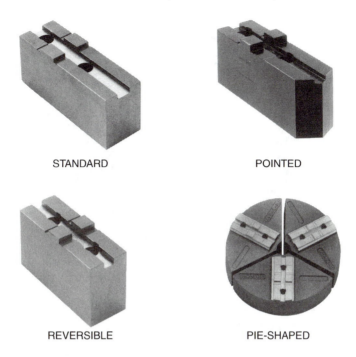

STANDARD          POINTED

REVERSIBLE        PIE-SHAPED

**Figure 9-2**. *Soft jaws are customized to fit a specific workpiece.*

Initially, for machining, the jaws are usually mounted on the chuck and clamped against a mandrel, or a ring for internal chucking. The jaws are then machined to the proper form and diameter for the workpiece. When machining chuck jaws, it is important to select the correct-size ring or mandrel to ensure the turned diameters on the jaws precisely match the workpiece.

Another way to machine soft jaws is with a top-jaw machining fixture, Figure 9-3. This fixture has a heat-treated scroll mechanism that advances or retracts prongs. The prongs engage the chuck jaws in their mounting holes. They also provide support for both internal and external jaw-turning operations. Because it eliminates jaw backlash and equalizes the clamping pressure between jaws, the fixture improves concentricity. The procedure to machine chuck jaws with this fixture is shown in Figure 9-4.

The selection of soft jaws is based on the type and size of chuck, jaw height, and jaw material. Another area that should also be considered is the jaw shape. Soft jaws are available in four basic shapes: standard,

**Figure 9-3.** *A top-jaw machining fixture allows more-accurate top-jaw machining. It improves concentricity by eliminating backlash and holding equal pressure on each jaw.*

STEP #1
Mount top jaws in
proper position for job.

STEP #2
Adjust fixture to allow inserting
prongs into top-jaw counterbores.

STEP #3
After inserting prongs,
actuate chuck.

STEP #4
Machine as required.

**Figure 9-4.** *The procedure for machining chuck jaws with a top-jaw machining fixture.*

pointed, reversible, and pie-shaped. The specific shape selected normally depends on the machining operations and the workpiece.

*Standard.* Standard soft jaws, Figure 9-5, are the most popular for general-purpose turning. These jaws are also called "offset" or "long" soft jaws because of their offset hole pattern. Soft jaws with the offset hole pattern are usually designed for non-reversible mounting. These jaws come in either standard or extra-high styles. After machining, they can be used in their soft condition for finished parts, or case hardened.

**Figure 9-5.** *Standard soft jaws.*

*Pointed.* Pointed soft jaws, Figure 9-6, are a variation of the standard soft jaws. Like standard jaws, these jaws also have an offset mounting-hole pattern. Pointed jaws work well for small-diameter workpieces because the pointed end allows the jaws to close nearer the center of the chuck. The standard included point angle on these jaws is 120°. Depending on the size of the jaws, they may also have a slight flat at the tip.

**Figure 9-6.** *Pointed soft jaws are ideal for small-diameter parts.*

*Reversible*. Reversible soft jaws, Figure 9-7, have the same basic shape as the standard soft jaws, but the mounting-hole pattern is in the center of the jaw rather than offset. These jaws are typically chosen for larger parts that require less removal of jaw material. Because of the centered mounting holes, the jaws can be machined on both sides and reversed for different chucking operations.

**Figure 9-7.** *Reversible soft jaws can be machined on both ends.*

*Pie-Shaped*. Pie-shaped jaws, Figure 9-8, are also individual jaws that are mounted directly to the master jaws on the chuck. Like other soft jaws, these chuck jaws are also furnished in a partially machined condition and are available in a variety of materials. The main difference between these jaws and the other soft jaws is their size: pie-shaped jaws are larger and generally cover the entire face of the chuck. The full-grip design allows increased surface contact on large-diameter or thinwall parts, Figure 9-9. This helps eliminate distortion or marring of the workpieces.

**Figure 9-8.** *Pie-shaped soft jaws (set of three).*

**Figure 9-9**. *The full-grip design of pie-shaped jaws allows increased surface contact on large-diameter or thinwall parts.*

Pie-shaped jaws also work well for parts with odd, non-cylindrical shapes, or parts which must have maximum support. In the case of odd or irregular parts, Figure 9-10, the jaws can be machined on a milling machine rather than a lathe. When clamping parts in pie-shaped jaws, always make sure a small space is between the jaws when the part is clamped. This ensures that the jaws are clamped against the part and not against each other. Shims should be placed between the jaws during machining to provide adequate allowance for part variations.

**Figure 9-10**. *Pie-shaped jaws are also well suited for parts with odd, non-cylindrical shapes.*

*Gripper Inserts.* Serrated round grippers, Figure 9-11, can be installed in customized chuck jaws to increase holding force. These hardened inserts are available in many types and sizes, in high-speed tool steel, carbide-tipped tool steel, and solid carbide.

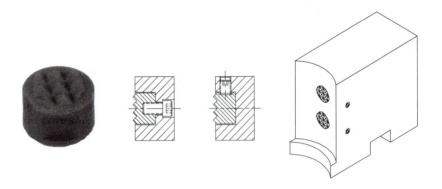

**Figure 9-11**. *Machined blank chuck jaws with serrated round grippers installed can provide holding force comparable to fully serrated hard jaws.*

## COLLET VISES

Collets and collet vises are another popular form of general-purpose workholder. Like chucks, collets and collet vises have many design advantages that make them very attractive and cost-effective workholders.

### Types of Collet Vises

Collet vises come in two general forms, single collet vises and triple collet vises. Both styles of collet vises are hydraulically operated and provide adaptable ways to clamp collet-mounted workpieces. Each of these vises can be used with any standard or special 5C collet. This includes standard collets, step collets, expanding collets, and emergency collets. Both vises can be set up individually, or gang mounted for multiple-workpiece machining setups.

*Single Collet Vises.* The single collet vise, Figure 9-12, is designed for single-part applications where only one collet is needed. These collet vises have two clamping cylinders, one on each side of the collet, to provide a smooth and uniform clamping motion. When the power is released, a dual-spring return mechanism releases the collet. Mounting holes in the vise body permit the single collet vise to be mounted either vertically or horizontally, as needed. An adjustment-screw plug is provided in the base of the vise, Figure 9-13. This plug permits the height of shorter workpieces to be precisely adjusted and set. For applications with longer parts, or for bar feeding, the adjustment-screw plug can be removed, Figure 9-14.

**Figure 9-12.** *The single collet vise is an accurate and versatile fixture for round, square, and hex-shaped parts.*

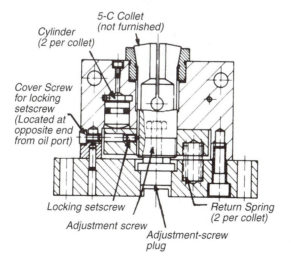

**Figure 9-13.** *The adjustment-screw plug in the base of a collet vise permits workpiece height to be precisely adjusted and set.*

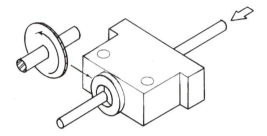

**Figure 9-14.** *The adjustment-screw plug can be removed for longer parts or bar-feeding operations.*

*Triple Collet Vises.* The triple collet vise, Figure 9-15, is designed for multiple-workpiece applications. These vises, like the single collet vise, have two clamping cylinders and two return springs for each collet. Like the single collet vise, the triple collet vise also has a removable adjustment-screw plug for each collet in the base of the vise. Both the single and triple collet vises can be mounted together to accommodate multiple-workpiece setups, as shown in Figure 9-16.

**Figure 9-15.** *The triple collet vise.*

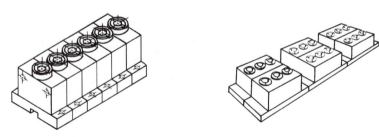

**Figure 9-16.** *Both the single and triple collet vises can be gang mounted for multiple workpiece setups.*

## Standard Collets

There are several types of collets available for workholding. Of these, the most popular is the 5C style, Figure 9-17. These collets are made in a variety of forms, but the standard round, square and hexagonal are the most common. Standard 5C collets are well suited for workholding applications. Since they are standard, off-the-shelf items, no machining is required.

**Figure 9-17.** *5C-style collet.*

*Round.* Round 5C collets, Figure 9-18a, are the most-popular collet. These collets are for round workpieces or barstock. Round 5C collets are made in standard sizes ranging from 1/64" to 1-1/16" in 1/64" increments.

*Square.* Square 5C collets, Figure 9-18b, are a variation of the round collet. They are designed for square workpieces or barstock. Square 5C collets are made in standard sizes ranging from 3/64" to 3/4" in 1/64" increments.

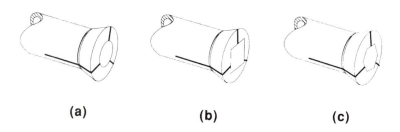

**(a)**　　　　　**(b)**　　　　　**(c)**

**Figure 9-18.** *Round, square and hexagonal 5C-style collets.*

*Hexagonal.* Hexagonal 5C collets, Figure 9-18c, are for holding hex-shaped workpieces and barstock. Hexagonal 5C collets are made in standard sizes ranging from 1/16" to 7/8" in 1/64" increments.

## Step Collets

Step collets, Figure 9-19, are an enlarged version of the standard 5C collet. Step chucks have a shallow gripping capacity, so they are most commonly used for thinner workpieces with larger diameters. These collets are available in sizes ranging up to 6.00" in diameter.

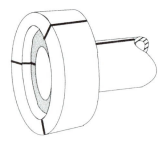

**Figure 9-19.** *Step collets are available for larger diameters.*

## Expanding Collets

Expanding collets, Figure 9-20, are a unique-but-popular collet. Unlike conventional collets that grip a workpiece on the outside and squeeze inward, these collets grip the workpiece in an internal bore and push outward. The machine-mounting end of the collet is the same shape as that for other 5C collets. Expanding collets are used to internally

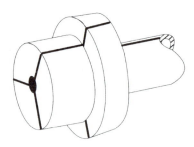

**Figure 9-20.** *Expanding collets are used for clamping an internal diameter.*

hold parts with diameters up to 6.00". These collets have less gripping force than conventional collets, so they are best suited for medium-duty clamping.

## Emergency Collets

Emergency collets are another 5C-collet variation. Emergency collets are made with a standard collet mount that is hardened and ground to size. The opposite end of the collet is soft and can be machined to meet individual workpiece requirements. Emergency collets are available in two basic styles, collet and step collet. The collet style, Figure 9-21, externally holds smaller parts and the step collet style, Figure 9-22, is for parts to 6.00" in diameter. The collet type can be used for longer parts, providing the diameter is smaller than 1.063". The step-collet type is for larger-diameter parts that are relatively thin, .50" or less.

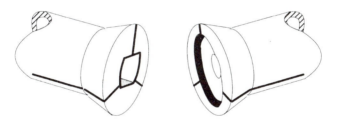

**Figure 9-21.** *Emergency collets can be customized for clamping odd shapes.*

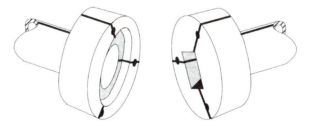

**Figure 9-22.** *Step-type emergency collets hold larger odd-shaped parts.*

These collets must first be machined to size. The collet is mounted in a lathe, or other machine tool, and machined to the desired form. Each of these collets is furnished with three pins between the gripping segments in the soft end, Figure 9-23. The pins must be in place while machining the collets to size. Once the collet is machined, the pins are

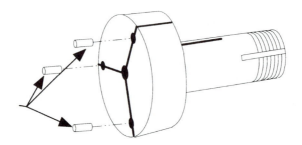

**Figure 9-23.** *Each emergency collet is furnished with three pins to keep the gripping segments properly spaced and rigid during machining.*

removed. For most applications these collets are used in their as-machined condition. For longer production runs, they occasionally require hardening.

## MACHINE VISES AND VISE JAWS

Machine vises are among the most-common workholders used in manufacturing today. Their main benefits include self-contained work-holding capability, a relatively fast clamp/unclamp cycle, accurate machine-table mounting, a solid fixed jaw for part location, and good rigidity.

### Machine-Vise Operation

Although manual vises are still frequently used in some areas of manufacturing, power vises are usually the best choice for high-production machine tools. There are several different types of power actuation, ranging from power assist to fully automatic operation, as shown in Figures 9-24 through 9-26. The following are the basic methods of vise operation.

*Manual.* Manual vises are found in almost every shop, in many varieties and sizes. Manual vises are the most-common and usually the cheapest type of vise. These vises are hand operated using a crank handle or similar device. Usually a simple lead screw advances the sliding jaw, clamping against the fixed jaw.

*Hydraulic.* Hydraulic vises provide highly consistent clamping force. After initial adjustment to the workpiece, these vises clamp automatically at the push of a button, powered by an external hydraulic

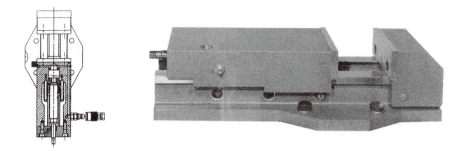

**Figure 9-24.** *Hydraulic vises provide automatic power clamping with exact high force at the push of a button (external power unit required).*

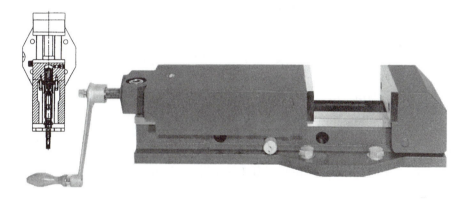

**Figure 9-25.** *Hydra-mechanical vises are manually operated, but final high force is applied easily using hydraulic power assist.*

**Figure 9-26.** *Hydra-pneumatic vises are also manually cranked, but final high force is applied by a built-in air-powered hydraulic intensifier.*

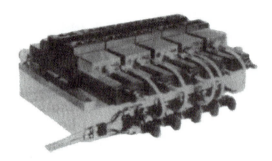

**Figure 9-27.** *Five hydraulic vises mounted side by side, operated independently by five clamping valves (using one hydraulic power unit).*

power unit. Hydraulic vises can be used individually or gang mounted for multiple-part setups, Figure 9-27, controlled individually by clamping valves.

*Hydra-Mechanical.* Hydra-mechanical vises are self-contained power vises that require no external hookup. Turning the crank handle clockwise moves the sliding jaw mechanically toward the workpiece. Once contact is made, further turning automatically disengages the spindle screw and gradually applies full hydraulic pressure, with relatively little effort. Markings on the spindle indicate the approximate percentage of clamping force attained.

*Hydra-Pneumatic.* Hydra-pneumatic vises have an air-powered hydraulic intensifier built into their sliding jaw (shop-air connection required). Turning the crank handle clockwise moves the sliding jaw mechanically toward the workpiece. Once contact is made, an additional 1/3 turn quickly applies full, high clamping force. Clamping force is set by adjusting input air pressure.

*Pneumatic.* Simple pneumatic vises are powered directly by air pressure, not hydraulic power. Since typical shop air systems operate at only 80-100 psi, large air cylinders are required, and still provide only light clamping force.

## Types of Machine Vises

There are countless varieties of vises, but the following are the major categories of machine vises used in production applications.

STANDARD VISE
WITH SWIVEL BASE

COMPOUND VISE

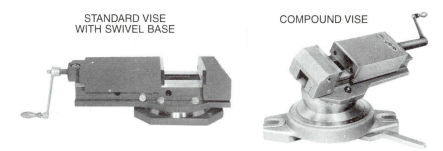

**Figure 9-28.** *Precision milling vises are usually mounted directly on a T-slotted machine table. Swivel bases and compound-angle bases are also available.*

*Precision Milling Vises.* This type of vise, Figure 9-28, is commonly used on vertical milling machines and other machine tools with T-slotted tables. These vises are slotted underneath for accurate two-axis positioning using fixture keys. Optional swivel bases and compound-angle bases can be used for machining at angles.

*High-Precision Machining-Center Vises.* These vises, Figure 9-29, have locating holes on two faces to allow mounting vertically or horizontally with excellent repeatability. They are specially designed to mount on horizontal-machining-center pallets, as well as vertical milling machines, providing accurate zero-position referencing in all three axes. Machining-center vises can be mounted in multiples on modular tooling blocks, with up to eight vises per pallet.

**Figure 9-29.** *High-precision machining-center vises have locating holes to allow mounting vertically or horizontally with excellent repeatability in all three axes. They are ideal for mounting in multiples on tooling blocks.*

**Figure 9-30**. *Double vises are high-precision vises that allow clamping two workpieces, against a fixed center locator, in about the same space as a normal vise.*

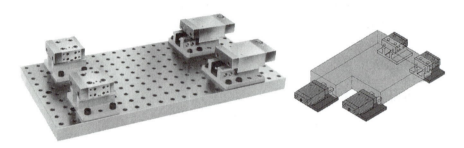

**Figure 9-31**. *Modular vises are high-precision vises consisting of self-contained fixed-jaw and sliding-jaw units that mount on modular tooling plates and blocks for total flexibility.*

*Double Vises.* These high-precision vises, Figure 9-30, allow clamping two workpieces, against a fixed center locator, in about the same space as a normal vise. This type features the same accurate location in all three axes as high-precision machining-center vises.

*Modular Vises.* These high-precision vises, Figure 9-31, consist of separate, self-contained fixed-jaw and sliding-jaw units. The jaw assemblies mount on modular tooling plates and blocks for accurate three-axis location.

## Vise Jaws

The standard jaws on most vises are plain hardened-and-ground flat jaws, sometimes with serrations for extra gripping force. Soft jaws can be machined to make custom vise-jaw fixtures. Optional vise jaws, Figure 9-32, are also available to expand a vise's versatility.

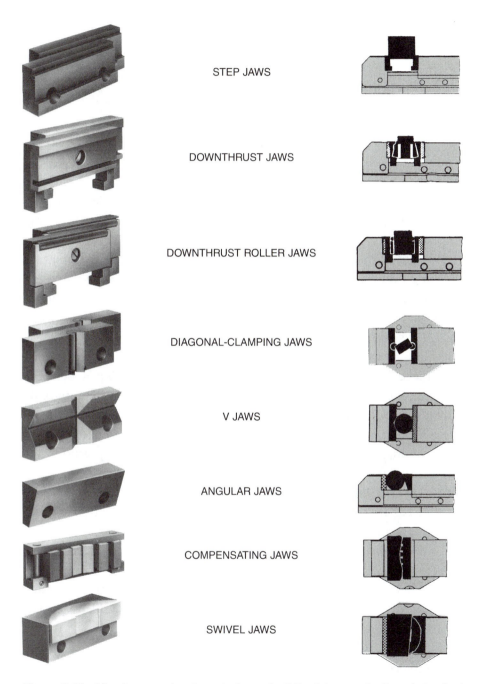

STEP JAWS

DOWNTHRUST JAWS

DOWNTHRUST ROLLER JAWS

DIAGONAL-CLAMPING JAWS

V JAWS

ANGULAR JAWS

COMPENSATING JAWS

SWIVEL JAWS

**Figure 9-32.** *Vise jaws are key to a vise's productivity. A large selection of standard jaws reduces the need to machine custom vise-jaw fixtures.*

*Step Jaws.* These jaws allow clamping a rectangular part high up in the vise without using parallel blocks. This raises short workpieces for easier access, and also provides tool clearance between the workpiece and vise bed. Each jaw has both a high and low step. Steps are precision ground to maintain accurate height and positional location. Step jaws should be used only in pairs.

*Downthrust Jaws.* These unique jaws push the workpiece downward while still providing accurate positional location. This feature is useful whenever extreme part-height accuracy is required. Each jaw's contact surface is ground flat to avoid marring finished surfaces. Downthrust jaws should be used only in pairs, leaving standard flat jaws in place.

*Downthrust Roller Jaws.* These downthrust jaws are designed for gripping unfinished surfaces. The pressure roller on top grips into the part for extra downthrust force. These jaws should be used only in pairs, leaving standard flat jaws in place.

*Diagonal-Clamping Jaws.* These unique jaws have a precision swiveling V groove. This allows accurately centering any rectangular workpiece between two opposite corners, while keeping it perfectly upright (eliminating the need for a side stop). Diagonal-clamping jaws should be used only in pairs.

*V Jaws.* This jaw has precisely ground V grooves to allow clamping cylindrical workpieces in either horizontal or vertical position. A V jaw is usually used together with a standard flat jaw.

*Angular Jaws.* This jaw allows clamping cylindrical and radiused workpieces firmly down against the vise bed. It should be used together with a standard flat jaw.

*Compensating Jaws.* This unique jaw has clamping pads that mechanically adjust to a contoured workpiece. Each pad applies approximately the same clamping pressure. A compensating jaw should be used only on the sliding-jaw side, together with a standard flat jaw on the fixed-jaw side (for positional accuracy).

*Swivel Jaws.* This jaw has a swiveling face for securely clamping a workpiece that has nonparallel sides. The contact face swivels up to 5° in both directions. A swivel jaw should be used together with a standard flat jaw.

**Figure 9-33.** *Self-centering vises have V jaws that self-center with highly accurate repeatability.*

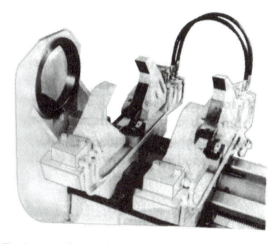

**Figure 9-34.** *Two large self-centering vises on a gun-drilling machine hold a workpiece for drilling.*

## SELF-CENTERING VISES

Self-centering vises, Figure 9-33, are hydraulically powered vises that combine a V-shaped-jaw design with a precision rack-and-pinion mechanism to provide highly accurate and repeatable clamping action. Self-centering vises are usually used in pairs, such as when clamping two ends of a shaft. They are available for workpieces from 3/8" to 24" in diameter. Figure 9-34 shows two self-centering vises used with a gun-drilling machine to hold the workpiece for a drilling operation.

# 10

# MODULAR FIXTURING

Jigs and fixtures have long been a part of manufacturing. But beyond their benefits, conventional jigs and fixtures also have a few shortcomings. This is especially true when production runs are small and do not repeat on a regular basis. When a company adds modular fixturing to its options, even job-shop-type production can benefit from quality fixturing.

## MODULAR FIXTURING'S ROLE IN WORKHOLDING

Modular fixturing is not intended for every workholding operation, but when it is appropriate, it both increases production and reduces fixturing costs. Modular fixturing is not a replacement for permanent fixturing; rather it is an upgrade from "no fixturing," just a step below permanent fixturing.

### Definition of Modular Fixturing

Modular fixturing is a workholding system using a series of standardized components for building specialized workholders. As shown in Figure 10-1, a modular workholder is assembled from a variety of standard off-the-shelf tooling plates, supports, locating elements, clamps, and similar components. The components are assembled with socket-head cap screws and locating screws. A modular-fixturing system may have hundreds of different elements. The components can be assembled in different combinations to build an unlimited variety of workholders.

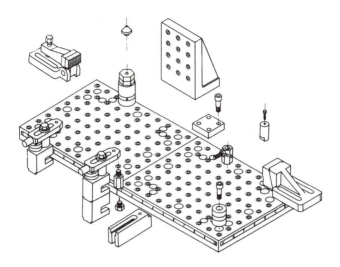

**Figure 10-1.** *Modular workholders can be assembled entirely from standard off-the-shelf components.*

The assembly process is quite simple. Components are designed to be used together, so each has an identical hole pattern. To build a workholder, the components are simply positioned as required and attached with locating or fastening screws. The simplicity reduces training time and permits technicians to begin building workholders almost immediately. Regardless of the manufacturing operations, a modular-fixturing system can provide workholders for almost any workpiece.

## The Hierarchy of Workholding Options

To understand how modular fixtures relate to workholding in general, an awareness of the various forms of workholders is necessary. Workholder forms can be grouped into three general categories, Figure 10-2. The least-complicated tools are the general-purpose workholders. The most-complex and most-detailed workholders are the special-purpose, or permanent, fixtures. Between the two are modular fixtures.

Modular fixturing bridges the gap between the general-purpose and special-purpose workholders. Although the three forms of fixturing may seem completely different, each is actually a further development, or refinement, in the workholding process.

General-purpose workholders are the simplest form of fixturing device. This category includes a wide variety of standard clamps, vises, chucks, and similar standard off-the-shelf components. General-

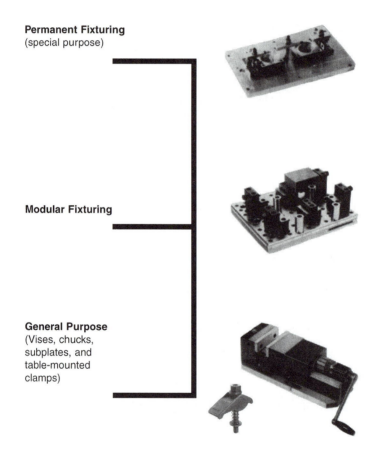

**Permanent Fixturing**
(special purpose)

**Modular Fixturing**

**General Purpose**
(Vises, chucks, subplates, and table-mounted clamps)

**Figure 10-2.** *The hierarchy of workholding options. Modular fixturing fills the significant gap between general-purpose and permanent workholders.*

purpose components are reusable and normally the least expensive. Although these components represent a smaller initial investment, they are often inadequate or unsatisfactory for complex parts or high-volume production.

Permanent workholders are specifically designed and constructed for a single workpiece or family of parts. These workholders, though usually the most efficient, are also the most expensive. Permanent workholders are built with a variety of standard and custom-made parts to meet specific requirements. These fixtures are the best choice for high-volume or repeated production runs.

Modular workholders, in the most-basic sense, can be described as special-purpose workholders assembled from general-purpose components. Figure 10-3 shows a typical system of modular-fixturing components. The concept of modular fixturing, rather than a departure from conventional fixturing methods, is actually a combination of the best attributes of both special- and general-purpose workholding methods. Modular fixtures are built with the accuracy and detail of special-purpose

**Figure 10-3.** *Typical components of a modular-workholding system.*

workholders, but with reusable and universal components, these fixtures compare favorably in cost to general-purpose workholders.

## Good Applications for Modular Fixturing

Modular workholders are particularly well suited for one-time jobs, infrequent productions runs, prototype parts, replacement parts, trial fixturing, and temporary tooling.

*One-Time Jobs.* One-time jobs, found especially in job shops, are ideal for modular fixturing. With modular fixturing, a workholder can be economically built even for a one-part run. In most job shops, each machine tool does a variety of tasks in a single day. With modular fixturing, jobs can be performed with specialized workholders at a cost very competitive with crude machine-table setups. Here, modular fixtures increase quality and accuracy, yet still maintain competitive costs.

*Infrequent Production Runs.* Jobs that do not repeat on a regular basis are well suited to modular fixturing. Today shorter lead times are quite common. Modular fixturing permits rapid setup of short-notice production runs. Once again, modular workholders offer many of the benefits of special-purpose tooling at a fraction of the cost.

*Prototype Parts.* Prototype or experimental parts frequently require special workholders. Since prototype workpieces are often changed or redesigned, the cost of building jigs or fixtures for each new variation is prohibitive. Modular fixturing is the best alternative. With modular workholders, each variation of the workpiece can be quickly fixtured with little or no downtime.

*Replacement Parts That Are Made to Order.* Replacement parts are an expensive problem in many companies. In the past, these parts were made in large lots and placed in storage. Some parts were quickly sold while others were never ordered. Modular fixturing eliminates the need for an inventory of slow-moving replacement parts. Modular fixturing permits a company to respond to orders as they are received. Instead of shipping parts from an inventory, parts can be made as needed.

*Trial Fixturing Techniques.* Trial fixturing is common throughout manufacturing organizations. Before any product goes into production, workholders must be designed, built, and tested. Assembling a modular workholder for workpiece allows the tool designer to test new tool designs and find problem areas. Modular workholders allow a designer to fine tune tooling ideas before a final production workholder is built.

*While Permanent Fixtures Are Built or Repaired.* No matter how well a permanent jig or fixture is built, most require repair or maintenance. A modular workholder can easily be assembled and placed into production while the permanent workholder is repaired. Modular fixtures also buy time for initial building of permanent fixtures without delaying production.

## Poor Applications for Modular Fixturing

Just as conventional machine tools are better suited for some tasks than expensive CNC machines, the selection and application of workholders is determined by the work to be performed. Modular workholders are not intended for every job. Two limiting factors with modular workholders are the frequency of production runs and the size of the workholder.

*Jobs That Will Repeat Many Times.* Recurring production runs are those that repeat on a regular basis. The choice of fixturing method depends more on the frequency of production runs than on the number of parts per run. As shown in Figure 10-4, choosing between modular and permanent fixturing for a particular job depends on how often the job will run, not just on lot size.

Modular workholders are usually disassembled after each job. Each time a job is run, the modular fixture requires reassembly. Permanent workholders are normally built for a complete product run. If a job repeats on a regular basis, a permanent fixture is the better choice. Modular fixtures could remain assembled between production runs, thus becoming a permanent workholder, but this negates the economy of reusing modular components.

*Where Fixture Compactness Is Important.* Another factor to consider in the selection of a workholder is the size of the completed fixture.

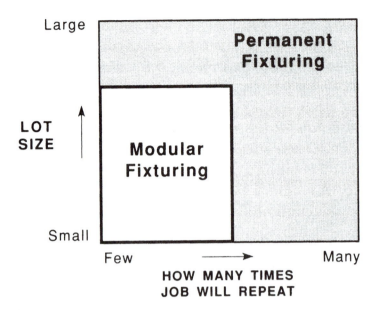

**Figure 10-4.** *Choosing between modular and permanent fixturing depends mainly on how often a job will run, not just on lot size.*

As a rule, modular workholders tend to be larger than their special-purpose counterparts. Permanent workholders are normally built from custom-made elements and baseplates that permit a smaller, more-compact workholder. Modular components are intended for a variety of applications and, though more universal, are larger than comparable custom-made components. So, when space is limited, such as with multiple-part setups, modular workholders may prove to be too large.

## IMPORTANT FEATURES OF A GOOD MODULAR-FIXTURING SYSTEM

The earliest modular-fixturing systems date back to early 1940's England, where tool designers continually needed to replace tooling destroyed by German bombs during World War II. Although pioneered decades ago, modular-fixturing concepts have only recently been perfected. Today, several different systems are commercially available, usually based on tooling plates and blocks with mounting holes arranged in a grid pattern, or, less commonly, with crossing T Slots. Following are some key features to look for when choosing a modular-fixturing system.

## A Wide Selection of Tooling Plates and Blocks

Tooling plates and blocks are the main structural elements of any modular-fixturing system. The type, style, and number of plates and blocks available determine the variety of machine tools that can be served with the system. To provide maximum versatility, the ideal system must offer a wide range of different base elements, Figure 10-5. Not only is a variety of styles and shapes important, but there should also be a selection of different sizes within each basic style. Although a wide selection of all components is important, a full range of available tooling plates and blocks increases a modular system's value.

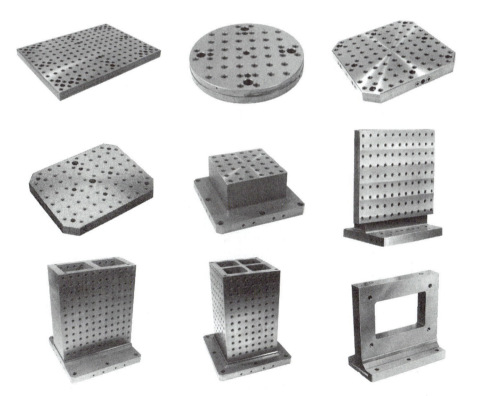

**Figure 10-5.** *A wide selection of available tooling plates and blocks is an important feature of a good modular-fixturing system.*

## Grid Holes vs. T Slots

The two primary forms of modular systems available today are those with grid-pattern holes and those with T slots to mount components, Figure 10-6. Grid holes offer several advantages over T slots. The

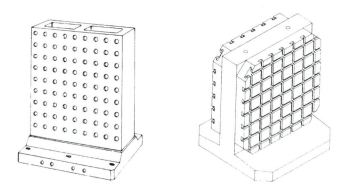

**Figure 10-6.** *Modular systems with grid holes have greater positional accuracy and strength than those with T slots.*

T slots are generally spaced further apart and offer fewer mounting positions. The grid holes offer many more positional possibilities. Even though T-slot system permit movement along the slot, nothing but friction holds components in place. Sudden jarring or excessive cutting force applied to components can cause movement.

The only places where T-slot systems have fixed reference points are at the intersections of the slots. Here too, since there are only a few slots, T slots offer fewer locational possibilities. The grid pattern offers more security since each hole is itself a locating point. Since these points are always fixed, repeated setups are easier to assemble.

The final difference between the two systems deals with strength. T slots are individually very strong, but wherever T slots cross each other, a weak point develops. When a component is located at the point where two T slots cross, excessive forces cause problems. Also, if grid-

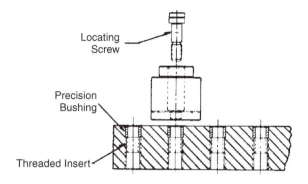

**Figure 10-7.** *A multipurpose hole has both a precision alignment bushing and a thread insert in the same hole. This allows mounting either a locator or a clamp in any hole.*

pattern holes are ever damaged, the damaged hole is repairable. T-slot plates and blocks, on the other hand, must be replaced when damaged.

## Multipurpose Holes

Grid-pattern modular-fixturing systems come in two styles: those with alternating dowel holes and tapped holes and those with multi-purpose holes. The alternating-hole style, though better than a T-slot system, does not offer as many advantages as the multipurpose-hole style. As shown in Figure 10-7, the multipurpose hole has both an alignment bushing and thread insert in the same hole. This arrangement permits each hole to act as an alignment hole, mounting hole, or both.

Multipurpose holes are laid out in a standard grid pattern to allow sufficient flexibility and still maintain ample plate strength.

## Choice of Several System Sizes

A modular fixturing system needs to accommodate a wide range of possible workpiece sizes, and many different machine tools. Considering the diversity of parts machined in different industries, a choice of several system sizes should be available. Larger parts are typically machined on larger, more-powerful machine tools to remove material at a faster rate, requiring greater clamping force. Small parts are often machined in multiples for greater productivity, and might be damaged by too much clamping force.

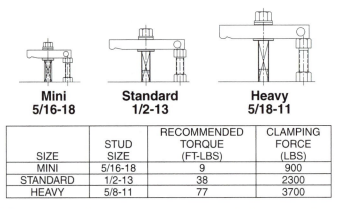

| SIZE | STUD SIZE | RECOMMENDED TORQUE (FT-LBS) | CLAMPING FORCE (LBS) |
|---|---|---|---|
| MINI | 5/16-18 | 9 | 900 |
| STANDARD | 1/2-13 | 38 | 2300 |
| HEAVY | 5/8-11 | 77 | 3700 |

**Figure 10-8.** *A good modular fixturing system is available in several system sizes. This table shows the typical clamping forces exerted by clamp straps in three different sizes.*

The three system sizes shown in Figure 10-8 cover almost any workpiece encountered in typical machine shops. Clamp size and force increase with system size. Hole spacing is tighter in small systems, Figure 10-9, to allow mounting more parts on a fixture. Mounting-hole dimensions also vary in proportion to system size, as shown in Figure 10-10.

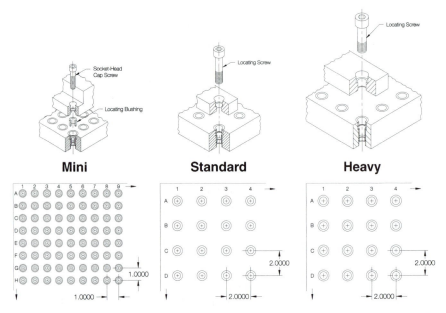

**Figure 10-9.** *Grid holes are more closely spaced in a small-size modular system. This allows locating small parts close together for machining in multiples.*

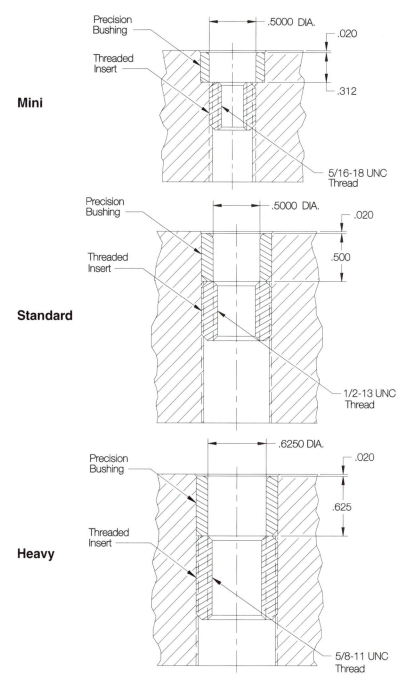

**Mini**

Precision Bushing

Threaded Insert

.5000 DIA.

.020

.312

5/16-18 UNC Thread

**Standard**

Precision Bushing

Threaded Insert

.5000 DIA.

.020

.500

1/2-13 UNC Thread

**Heavy**

Precision Bushing

Threaded Insert

.6250 DIA.

.020

.625

5/8-11 UNC Thread

**Figure 10-10.** *These cross sections show multipurpose holes in three different system sizes (actual size).*

By carefully selecting the most-useful system size based on typical parts and machine tools, a manufacturer can cover most of his fixturing requirements. Sometimes, using two system sizes is the best choice.

## Power-Workholding Capability

The ability to use power workholding is another consideration in the selection of a modular-workholding system. A modular system should treat power workholding as an integral part of the complete fixturing plan. Instead of requiring special power components, good modular systems use standard off-the-shelf power-workholding components. Adapting standard power components to modular workholders both reduces the cost and expands the design options. Figure 10-11 shows a modular workholder assembled with standard power-workholding components. The standard modular elements are combined with fluid-advanced work supports and swing clamps. The power components are attached to the modular baseplate with adaptor plates, then connected with fixture hoses.

**Figure 10-11.** *Standard power-workholding components can be used on modular workholders.*

## TOOLING PLATES AND BLOCKS

Tooling plates and blocks are the main structural elements of any modular-fixturing system. These elements are available in several forms for maximum design flexibility. In addition, each tooling plate and block variation comes in a variety of sizes for a wide range of applications. The following describes the principal types of tooling plates and blocks.

## Rectangular Tooling Plates

Rectangular tooling plates are the most-common base elements for modular workholders. As shown in Figure 10-12a, these plates have grid multipurpose-hole locations. The grid provides a precise and accurate locational base for other modular components in a typical workholder. The address numbers on two sides help identify specific grid locations. For long machine tables, two or more rectangular tooling plates can be combined using locating gages, Figure 10-12b. These alignment gages are removed once plates are fastened to the machine table.

## Round Tooling Plates

Round tooling plates, Figure 10-12c, are often used when a round base element is more appropriate for the workholder. Round tooling plates are used for rotary tables or round machine tables. These plates also have a grid pattern of multipurpose holes for mounting components.

## Machining-Center Pallets

Machining-center pallets are a group of modular base elements for palletized operations on machining centers. These tooling plates and blocks also have a standard grid pattern of multipurpose holes for mounting components. As shown in Figure 10-13, these plates and blocks come in several different forms. Most types are available to fit standard pallet sizes of 320mm, 400mm, 500mm, 630mm, and 800mm.

*Square Pallet Tooling Plates.* Square tooling plates, shown at (a), are the most-common tooling plates for pallet setups. These tooling

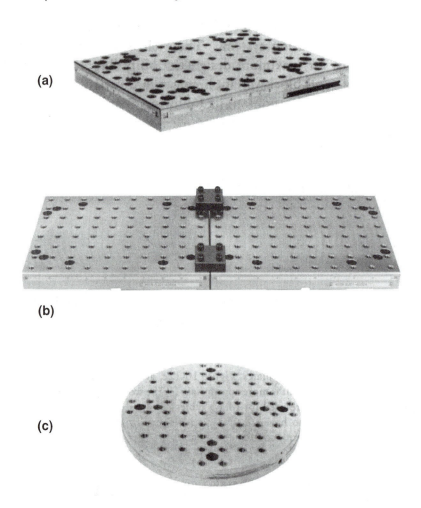

**Figure 10-12.** *Rectangular and round tooling plates.*

plates are also good for other setups that require a square base shape.

*Rectangular Pallet Tooling Plates.* These plates, shown at (b), are similar to the square tooling plates above, except that they fit standard rectangular pallets of 320 x 400mm, 400 x 500mm, 500 x 630mm, and 630 x 800mm.

*Platform Tooling Plates.* The platform tooling plate, shown at (c), is a modification of the square pallet tooling plate. These plates have an elevated mounting surface designed for horizontal machining centers.

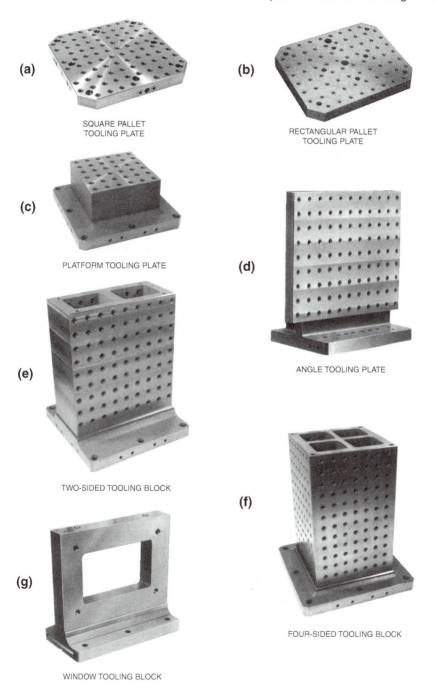

(a)

SQUARE PALLET
TOOLING PLATE

(b)

RECTANGULAR PALLET
TOOLING PLATE

(c)

PLATFORM TOOLING PLATE

(d)

ANGLE TOOLING PLATE

(e)

TWO-SIDED TOOLING BLOCK

(f)

FOUR-SIDED TOOLING BLOCK

(g)

WINDOW TOOLING BLOCK

**Figure 10-13.** *Machining-center pallets.*

The elevated area provides clearance for the machine spindle. It also allows rigid mounting of a workpiece that is elevated above the machine table.

*Angle Tooling Plates.* This plate, shown at (d), is a single vertical tooling surface perpendicular to the machine table. Gussets behind the plate provide excellent rigidity while keeping weight to a minimum.

*Two-Sided Tooling Blocks.* Two-sided tooling blocks are used for vertically mounted workpieces. As shown at (e), these tooling blocks have two mounting faces. Here, a workpiece is mounted on either one or both sides for machining. The two large faces are ideal for holding large workpieces.

*Four-Sided Tooling Blocks.* The four-sided tooling block, shown at (f), is also for vertically mounted workpieces. These tooling blocks have four identical mounting faces. The blocks are often used for machining four identical workpieces, but can also be used for a different operation, or different workpiece, on each face.

*Window Tooling Blocks.* This two-sided tooling block, shown at (g), is a precision frame for mounting quick-change vertical tooling plates. This allows leaving the window tooling block permanently on the machine table, and only handling relatively light plates to change a fixture setup. Also, the center opening can be used to machine a workpiece from the back side.

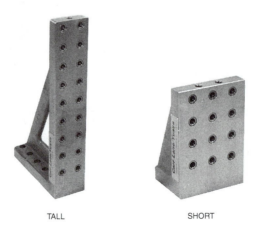

TALL                          SHORT

**Figure 10-14.** *Angle plates.*

## Angle Plates

Angle plates, Figure 10-14, are usually supplemental elements mounted on another style of tooling plate or block. These plates permit mounted elements to be set precisely at 90° to the base. The plates come in two styles, tall and short. Angle plates can be used for mounting workpieces directly, or as a base for attaching other components.

## Riser Plates

Riser plates, Figure 10-15, are another form of supplemental tooling plate. Elements can be mounted either vertically or horizontally on these plates. Riser plates elevate components to suit taller workpieces. Riser plates can also be used for directly mounting workpieces, or as a base for attaching other components.

**Figure 10-15.** *Riser plates.*

## MOUNTING ACCESSORIES

Mounting accessories are a group of components specifically designed for mounting the wide variety of modular components. The more-common mounting accessories include locating screws, riser blocks, riser cylinders, and adaptors.

## Locating Screws

Locating screws, Figure 10-16, are precision screws to both locate and mount modular components in the multipurpose holes. As shown, the screws are installed through the component and are aligned in the precision bushing. The threaded end securely attaches the component to the base element.

**Figure 10-16.** *Locating screws align and fasten simultaneously.*

## Riser Blocks

Riser blocks, Figure 10-17a, provide an elevated surface for mounting a clamp or locator. The blocks are made in several heights and have a series of threaded holes to position the mounted element precisely where it is required.

## Riser Cylinders

Riser cylinders, Figure 10-17b, like riser blocks, provide an elevated mount for a clamp or locator. The cylinders are used for elements with a single mounting hole. They work well where space is limited. Riser cylinders come in a variety of heights.

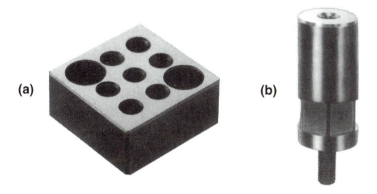

**(a)**　　　　**(b)**

**Figure 10-17.** *Riser blocks and riser cylinders are used to elevate components.*

## Adaptors

Adaptors are another valuable addition to a modular-fixturing system. Adaptors permit standard jig-and-fixture components to be used with the modular elements. Locating-pin adaptors, Figure 10-18a, are an example. They adapt standard round and diamond pins for use with modular elements.

Spring-stop-button adaptors, Figure 10-18b, are another useful adaptor. These units permit standard spring stop buttons and spring locating pins to be used with a modular workholder. A single adaptor works with a variety of spring stop buttons and spring locating pins.

Another adaptor that finds many applications in modular workholders is the toggle-clamp adaptor, Figure 10-18c. Toggle-clamp adaptors are attached to the modular components with a center mounting hole. The toggle clamps are then mounted to the adaptor. This adaptor permits a wide variety of toggle clamps to be used in a modular workholder.

## LOCATORS

Locators for modular-fixturing systems serve the same purpose as those for special-purpose fixtures. The only real difference between the two styles of locators is their application. While conventional locators are permanently installed in a single workholder, modular locators are reused many times on a variety of workholders.

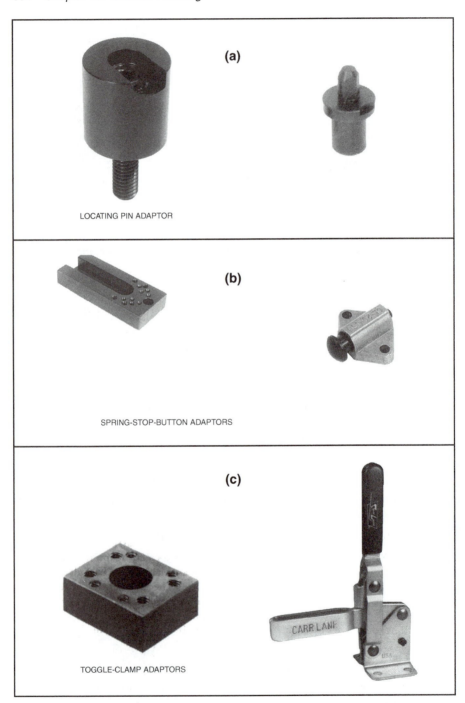

**Figure 10-18.** *Adaptors allow using standard components on modular fixtures.*

## Support Cylinders

Support cylinders are among the most-common locators for a modular system. As shown in Figure 10-19, support cylinders are available in three forms, plain, shoulder type and relieved. Plain support cylinders have a precisely ground cylindrical form, providing location on either their top or side. Shoulder support cylinders are made with a ground stepped diameter, providing location in both the vertical and horizontal axes. Relieved support cylinders are designed for supporting under a hole to be drilled.

PLAIN     SHOULDER TYPE     RELIEVED

**Figure 10-19.** *Support cylinders are used for both locating and supporting.*

## Screw Rest Pads

Screw rest pads, Figure 10-20, are solid rests often used along with other modular elements. Screw rest pads have a smaller contact area and provide a fixed locational point in areas where space is limited. These pads can be attached directly to the tooling plate or to another element such as the extension support shown.

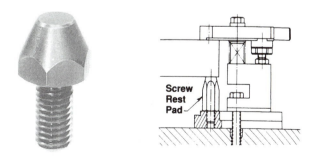

Screw Rest Pad

**Figure 10-20.** *Screw rest pads provide solid support with only a small contact area.*

## Round and Diamond Locating Pins

Round and diamond locating pins, Figure 10-21, are normally for permanent workholders, but with adaptors, these locating pins can also be used for modular applications. As shown at (b), fixed adaptors mount these pins directly to the holes in the tooling plates. Where locators are positioned between holes, the adjustable adaptors are used.

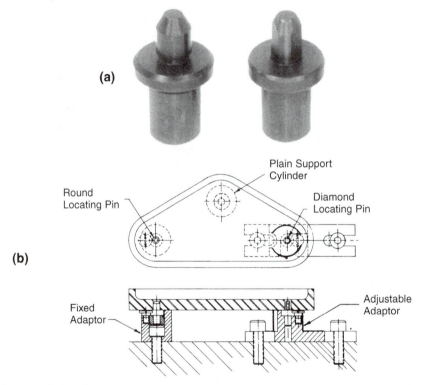

**Figure 10-21.** *Round and diamond locating pins are often used together to locate from holes in a workpiece.*

## Edge Supports

Edge supports, Figure 10-22, are another form of locator common in modular workholders. These supports come in two styles, single edge and double edge. The single edge supports locate single parts on a locating step, in both the horizontal and vertical axes. The double edge support serves an identical purpose, but instead of a single step, these supports have two steps, for mounting two different workpieces, one on each side.

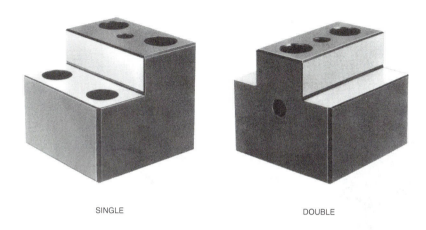

SINGLE

DOUBLE

**Figure 10-22.** *Edge supports provide support and side location simultaneously.*

## Extension Supports

Extension supports, Figure 10-23, are another unique locator for modular workholders. They are slotted locators with a precision-ground top surface and step. The slot permits the locator to be positioned at any point on the tooling plate. The step can locate a workpiece in both the horizontal and vertical axes. The height of the step matches the step height of the edge supports and other accessories. These supports can locate a workpiece directly, or hold additional elements installed in the tapped hole.

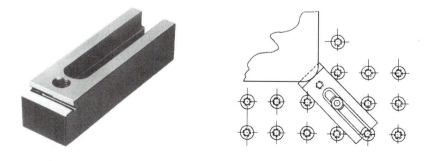

**Figure 10-23.** *Extension supports provide workpiece support between mounting holes.*

## Screw Jacks

Screw jacks, Figure 10-24, are heavy-duty supporting elements that provide adjustable height. They can be attached directly to the tooling plate or to other elements. A useful feature of these screw jacks is the interchangeable tips, which allow the contact area to be changed to suit the workpiece.

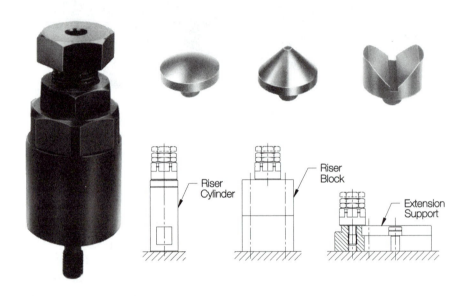

**Figure 10-24.** *Screw jack and interchangeable tips.*

## Manual Work Supports

Manual work supports, Figure 10-25, are another device suitable for modular fixturing. They have a slotted body for maximum positional flexibility. The support is positioned where needed and attached through the slot. The spring-loaded plunger floats until locked by turning the handle.

Manual work supports fill many support needs. They also work well as vibration dampeners. As shown at (b), a manual work support can be used in conjunction with a clamp. The clamp is attached to the tooling plate through the manual work support.

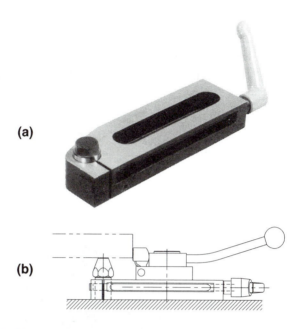

**(a)**

**(b)**

**Figure 10-25.** *Manual work supports can be positioned in hard-to-reach areas under the workpiece.*

## Adjustable Stops

Adjustable stops, Figure 10-26, act as adjustable locators. They come with a variety of holes for different workpiece heights. To change the height, the screw is simply removed from one hole and installed in another.

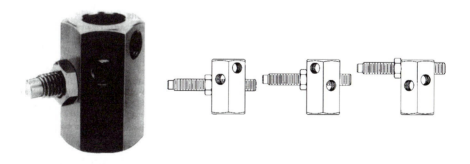

**Figure 10-26.** *Adjustable stops have an adjustable contact screw that can be mounted at different heights.*

## V Blocks

V blocks are an integral part of many workholders. The V blocks for modular fixtures come in two styles, horizontal and vertical, Figure 10-27. Both styles of V blocks are mounted with two mounting holes. Though identified as horizontal and vertical, V blocks can be used in any position depending on how they are mounted on the modular components. The horizontal V blocks offer more support area. The vertical style is best where space is limited.

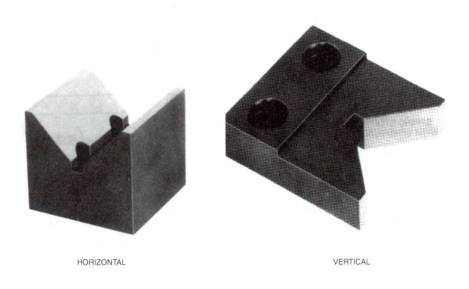

HORIZONTAL                               VERTICAL

**Figure 10-27.** *V blocks are available in horizontal or vertical styles.*

## Spring Stop Buttons

Spring stop buttons, Figure 10-28, are often used as a "third hand." They hold workpieces against the locators while the workpieces are clamped. These spring-loaded devices apply pressure against a workpiece to maintain proper contact against the locators. The buttons are mounted to the modular elements with an adaptor plate (a). A typical application with these buttons is shown at (b).

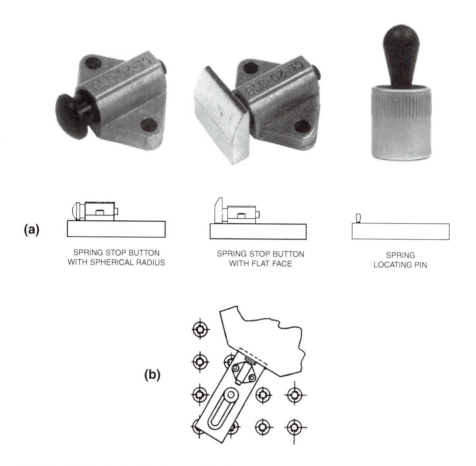

**Figure 10-28.** *Spring stop buttons hold a workpiece in position white it is clamped.*

## CLAMPS

Many types of downholding clamps and side clamps can be used for modular fixturing. Some are designed specifically for modular systems, but many are the same standard clamps used on permanent fixtures.

### Clamp-Strap Assemblies

Modular fixtures, like conventional fixtures, can also use clamp-strap assemblies to hold workpieces. As shown in Figure 10-29, slotted-heel and double-end clamp straps work well in a modular workholder.

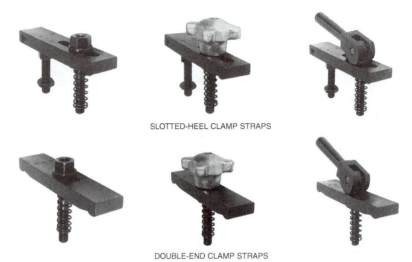

SLOTTED-HEEL CLAMP STRAPS

DOUBLE-END CLAMP STRAPS

**Figure 10-29.** *Almost any clamp-strap variation can be used on a modular workholder.*

## High-Rise Clamps

High-rise clamps, Figure 10-30, consist of a series of riser elements, special contacts, and clamp straps that can be combined in a variety of forms. These clamps hold a wide variety of workpiece shapes. The design of high-rise clamps permits the workpiece to be clamped either directly on the base element or elevated off the base with clamping accessories.

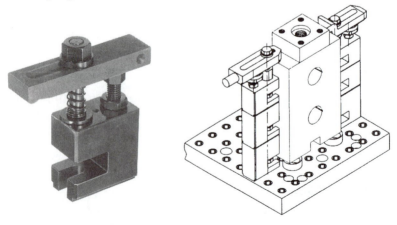

**Figure 10-30.** *High-rise clamps are all-in-one clamp-strap assemblies ideal for clamping tall workpieces.*

## Forged Adjustable Clamps

Another standard clamp for modular-fixturing applications is the forged adjustable clamp, Figure 10-31. These clamps act much the same way as clamp-strap arrangements, but they do not require a separate heel support. The shape of forged adjustable clamps permits them to clamp a range of different-size workpieces.

**Figure 10-31.** *Forged adjustable clamps have a built-in heel support.*

## Toggle Clamps

Toggle clamps are another clamp variety adapted for modular workholders. Several types of these clamps can be mounted to the modular components with either a fixed or slotted adaptor, Figure 10-32.

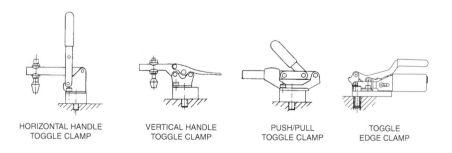

HORIZONTAL HANDLE
TOGGLE CLAMP

VERTICAL HANDLE
TOGGLE CLAMP

PUSH/PULL
TOGGLE CLAMP

TOGGLE
EDGE CLAMP

**Figure 10-32.** *Many types of toggle clamps can be mounted on modular fixtures using a simple fixed or slotted adaptor.*

## Up-Thrust Clamps

Up-thrust clamps, Figure 10-33, are a unique clamp design well suited for a variety of workholding tasks. These clamps have a cam-type clamp lever. The lever pulls the lower jaw upward. Since the workpiece is held with force from below, the clamps allow the workpiece to be located on its top surface rather than the bottom. This arrangement is ideal for workpieces with irregular bottom surfaces, such as the part shown.

The jaw element rotates. It has two clamp openings, one for thin parts and a second for thicker parts. In the example, the left side of the part is held with the smaller-opening jaw, while the right side is held with the larger-opening jaw. The underside of the top jaw element is a ground surface and acts as a precision locator for the clamped workpiece.

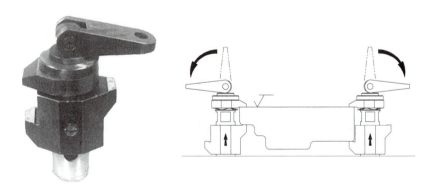

**Figure 10-33.** *Up-thrust clamps locate from a part's top surface and clamp from below.*

## Swing-Clamp Assemblies

Swing clamps, Figure 10-34, have a swinging clamp arm to speed clamping and unclamping. These clamps are available with either knob handles or ball handles. Swing clamps can be mounted directly to a tooling plate or on top of other modular elements. The clamp shown at (b) is mounted on riser blocks to gain additional height.

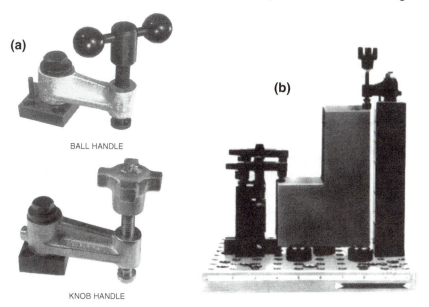

**Figure 10-34.** *Swing clamps are another standard clamping component for modular workholding.*

## Hook Clamps

Hook clamps, Figure 10-35, are a smaller variation of the swing clamp. Like swing clamps, the hook clamps for modular workholding come in a variety of sizes and two general styles, plain and tapped. A hook clamp holder, shown at (b), mounts these clamps to a modular workholder. When additional height is needed, a riser cylinder can be installed under the hook clamp holder.

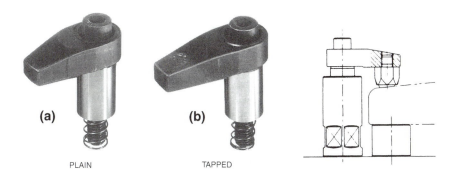

**Figure 10-35.** *Hook clamps and mounting adaptors.*

## Serrated Adjustable Clamps

The serrated adjustable clamp, Figure 10-36, is a standard clamp design also used for modular workholders. These clamps can be mounted directly to the tooling plate or on top of an extension support, as shown at (b). This clamp design transfers the thrust of an internal screw into clamping force, acting both forward and down along a 45° angle. The mounting slot has an angular bottom which prevents movement of the clamp body away from the workpiece.

These clamps are made with either a low-nose or high-nose design, shown at (c). Both types have gripping serrations on the clamping jaws. The high-nose clamp also comes with an aluminum cap that prevents serration damage to the workpiece.

**(a)**

**(b)**

**(c)**  Low Nose    High Nose

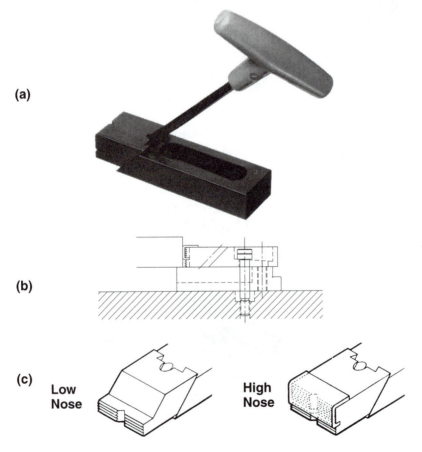

**Figure 10-36.** *Serrated adjustable clamps can be used on modular fixtures.*

## Cam Edge Clamps

The cam edge clamp, Figure 10-37, has a spiral-cam design. The cam moves the pivoting nose element and applies force both forward and downward to hold the workpiece. These clamps can be mounted directly to the tooling plate or on a riser block.

**Figure 10-37.** *Cam edge clamps have a slotted base for mounting flexibility.*

## Pivoting Edge Clamps

The pivoting edge clamp, Figure 10-38, is another side-clamping design. Instead of using an angular ramp to apply the holding force, these clamps generate force with a screw and pivot. As shown, these clamps are commonly used with a matching backstop to hold the workpiece between two jaw elements. Both the pivoting edge clamp and backstop have mounting slots and can be positioned anywhere on the workholder.

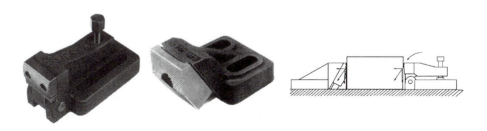

**Figure 10-38.** *Pivoting edge clamp and backstop.*

## Screw-Clamp Adaptors

Screw clamps, Figure 10-39, are another modular clamping component widely used for a variety of applications. The clamping screw is set at a slight downward angle and applies force both forward and downward against the workpiece. These adaptors can be used to mount hand-knob assemblies, swivel screw clamps, swivel head screws, and adjustable-torque thumb screws.

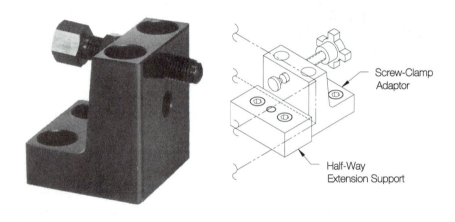

**Figure 10-39.** *Screw-clamp adaptors are another type of edge clamp.*

## POWER WORKHOLDING

Power workholding components are also used for many modular workholding applications. These components are mounted to the modular elements either directly, or with adaptors. Adapting standard power components to modular fixturing reduces tooling cost, because all components are reusable. When a power workholder is assembled on a modular base, the only additional elements required are the adaptors. These adaptors are grouped into three general categories: clamp adaptors, work-support adaptors, and valve and manifold adaptors. Precision vises are designed to mount directly, without adaptors.

## Clamp Adaptors

One clamp adaptor, Figure 10-40a, is used to mount extending clamps, edge clamps, or block clamps. This adaptor has a series of holes for mounting the clamps and a single mounting slot to attach the adaptor to the modular elements. As shown in Figure 10-40b, this slot allows the adaptor to be positioned in a variety of ways. Figure 10-40c shows how these adaptors work.

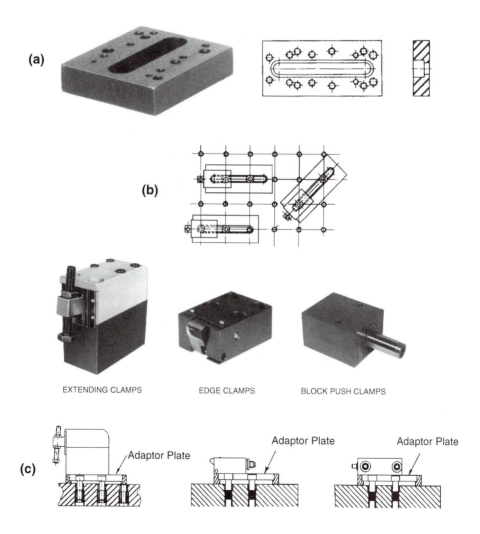

**(a)**

**(b)**

EXTENDING CLAMPS          EDGE CLAMPS          BLOCK PUSH CLAMPS

**(c)**          Adaptor Plate          Adaptor Plate          Adaptor Plate

**Figure 10-40.** *A single clamp adaptor can adapt several types and sizes of power clamps to modular fixtures.*

A second adaptor, shown in Figure 10-41a, is used for mounting swing clamps. This swing-clamp adaptor resembles the other clamp adaptor, but is wider to accommodate the swing-clamp base. This adaptor can mount two different sizes of low-profile or flange-base swing clamps, as shown in Figure 10-41b.

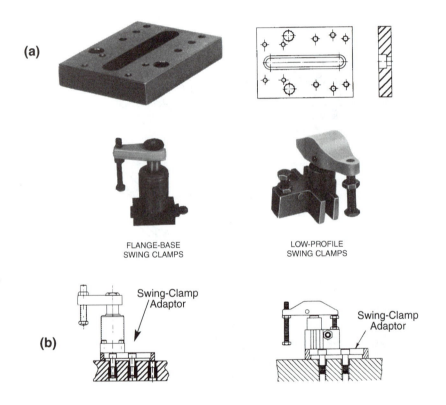

**Figure 10-41.** *A swing-clamp adaptor allows mounting several different types and sizes of swing clamps on modular fixtures.*

## Work-Support Adaptors

Work-support adaptors work with both the spring-extended and fluid-advanced work supports. As shown in Figure 10-42, there are three different styles of adaptors. The specific adaptor selected is determined by the type and size of work support used. The adaptor shown at (a) has the fixture-hose connection located at the end of the adaptor.

The adaptor at (b) has three holes for the fixture-hose connection. The hose may be positioned as shown, or at either 90° or 180° from this location. The larger adaptor, shown at (c), provides a mounting platform for the flange base work supports.

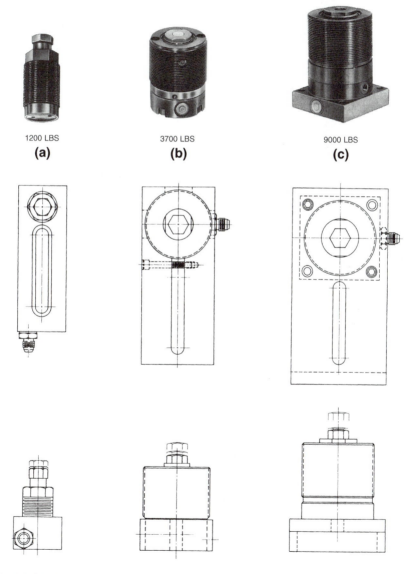

**Figure 10-42.** *Work supports can be easily mounted on modular fixtures, using work-support adaptors.*

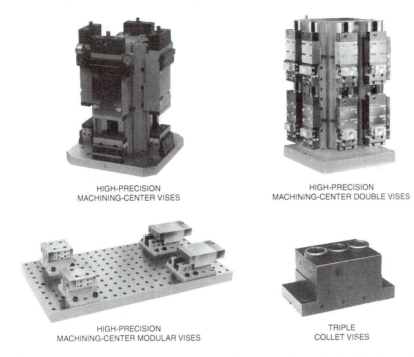

HIGH-PRECISION
MACHINING-CENTER VISES

HIGH-PRECISION
MACHINING-CENTER DOUBLE VISES

HIGH-PRECISION
MACHINING-CENTER MODULAR VISES

TRIPLE
COLLET VISES

**Figure 10-43.** *Precision vises can be easily mounted on standard tooling plates and blocks, with accurate three-axis position provided by locating holes.*

## Precision Vises

Several types of precision vises have locating holes for accurate mounting on modular tooling plates and blocks. This includes all high-precision machining-center vises, including double vises and modular vises, plus triple collet vises, as shown in Figure 10-43. Vises are mounted accurately in all three axes using two locating pins, one round and one diamond, Figure 10-44.

ROUND
LOCATING PIN

DIAMOND
LOCATING PIN

**Figure 10-44.** *One round and one diamond locating pin are used to locate all types of precision vises (and quick-change fixture plates) on modular tooling plates and blocks.*

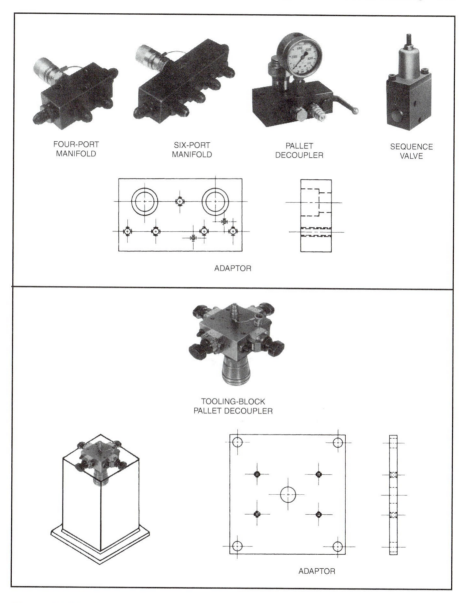

**FOUR-PORT MANIFOLD**

**SIX-PORT MANIFOLD**

**PALLET DECOUPLER**

**SEQUENCE VALVE**

ADAPTOR

**TOOLING-BLOCK PALLET DECOUPLER**

ADAPTOR

**Figure 10-45.** *Valve and manifold adaptors.*

## Valve and Manifold Adaptors

Valve and manifold adaptors securely attach a variety of sequence valves, manifolds, and pallet decouplers to a modular workholder, as shown in Figure 10-45.

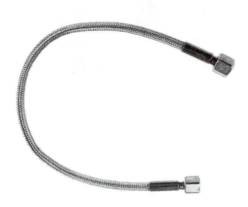

**Figure 10-46.** *Fixture hose.*

## Fixture Hose

Fixture hose, Figure 10-46, is a unique small-diameter, braided-steel-covered hydraulic hose that is ideally suited for modular-fixturing applications. This hose can be installed in the same tight-fit areas as custom-bent tubing, yet installs as easily as conventional hydraulic hose. Keeping a variety of different-length fixture hoses on hand ensures that virtually any modular power-workholding fixture can be assembled without custom fabrication.

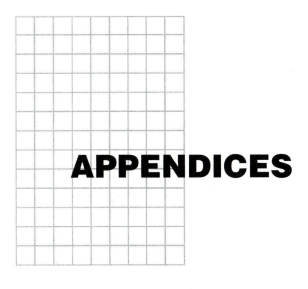

# APPENDICES

# *Carr Lane* DECIMAL EQUIVALENTS

| Fraction | Decimal | | Fraction | Decimal |
|---|---|---|---|---|
| $\frac{1}{64}$ | .015625 | | $\frac{33}{64}$ | .515625 |
| $\frac{1}{32}$ | .03125 | | $\frac{17}{32}$ | .53125 |
| $\frac{3}{64}$ | .046875 | | $\frac{35}{64}$ | .546875 |
| $\frac{1}{16}$ | .0625 | | $\frac{9}{16}$ | .5625 |
| $\frac{5}{64}$ | .078125 | | $\frac{37}{64}$ | .578125 |
| $\frac{3}{32}$ | .09375 | | $\frac{19}{32}$ | .59375 |
| $\frac{7}{64}$ | .109375 | | $\frac{39}{64}$ | .609375 |
| $\frac{1}{8}$ | .125 | | $\frac{5}{8}$ | .625 |
| $\frac{9}{64}$ | .140625 | | $\frac{41}{64}$ | .640625 |
| $\frac{5}{32}$ | .15625 | | $\frac{21}{32}$ | .65625 |
| $\frac{11}{64}$ | .171875 | | $\frac{43}{64}$ | .671875 |
| $\frac{3}{16}$ | .1875 | | $\frac{11}{16}$ | .6875 |
| $\frac{13}{64}$ | .203125 | | $\frac{45}{64}$ | .703125 |
| $\frac{7}{32}$ | .21875 | | $\frac{23}{32}$ | .71875 |
| $\frac{15}{64}$ | .234375 | | $\frac{47}{64}$ | .734375 |
| $\frac{1}{4}$ | .25 | | $\frac{3}{4}$ | .75 |
| $\frac{17}{64}$ | .265625 | | $\frac{49}{64}$ | .765625 |
| $\frac{9}{32}$ | .28125 | | $\frac{25}{32}$ | .78125 |
| $\frac{19}{64}$ | .296875 | | $\frac{51}{64}$ | .796875 |
| $\frac{5}{16}$ | .3125 | | $\frac{13}{16}$ | .8125 |
| $\frac{21}{64}$ | .328125 | | $\frac{53}{64}$ | .828125 |
| $\frac{11}{32}$ | .34375 | | $\frac{27}{32}$ | .84375 |
| $\frac{23}{64}$ | .359375 | | $\frac{55}{64}$ | .859375 |
| $\frac{3}{8}$ | .375 | | $\frac{7}{8}$ | .875 |
| $\frac{25}{64}$ | .390625 | | $\frac{57}{64}$ | .890625 |
| $\frac{13}{32}$ | .40625 | | $\frac{29}{32}$ | .90625 |
| $\frac{27}{64}$ | .421875 | | $\frac{59}{64}$ | .921875 |
| $\frac{7}{16}$ | .4375 | | $\frac{15}{16}$ | .9375 |
| $\frac{29}{64}$ | .453125 | | $\frac{61}{64}$ | .953125 |
| $\frac{15}{32}$ | .46875 | | $\frac{31}{32}$ | .96875 |
| $\frac{31}{64}$ | .484375 | | $\frac{63}{64}$ | .984375 |
| $\frac{1}{2}$ | .5 | | 1 | 1. |

# USA THREADS

| Thread Size* | Basic Major Diameter | Tap Drill Size** |
|---|---|---|
| #0-80 | .0600 | 3/64 |
| **#2-56** | .0860 | #50 |
| #2-64 | .0860 | #50 |
| **#4-40** | .1120 | #43 |
| #4-48 | .1120 | #42 |
| **#5-40** | .1250 | #38 |
| #5-44 | .1250 | #37 |
| **#6-32** | .1380 | #36 |
| #6-40 | .1380 | #33 |
| **#8-32** | .1640 | #29 |
| #8-36 | .1640 | #29 |
| **#10-24** | .1900 | #25 |
| #10-32 | .1900 | #21 |
| **1/4-20** | .2500 | #7 |
| 1/4-28 | .2500 | #3 |
| **5/16-18** | .3125 | F |
| 5/16-24 | .3125 | I |
| **3/8-16** | .3750 | 5/16 |
| 3/8-24 | .3750 | Q |
| **7/16-14** | .4375 | U |
| 7/16-20 | .4375 | 25/64 |
| **1/2-13** | .5000 | 27/64 |
| 1/2-20 | .5000 | 29/64 |
| **9/16-12** | .5625 | 31/64 |
| 9/16-18 | .5625 | 33/64 |
| **5/8-11** | .6250 | 17/32 |
| 5/8-18 | .6250 | 37/64 |
| **3/4-10** | .7500 | 21/32 |
| 3/4-16 | .7500 | 11/16 |
| **7/8-9** | .8750 | 49/64 |
| 7/8-14 | .8750 | 13/16 |
| **1-8** | 1.0000 | 7/8 |
| 1-14 | 1.0000 | 15/16 |
| **1-1/8-7** | 1.1250 | 63/64 |
| 1-1/8-12 | 1.1250 | 1-3/64 |

# METRIC THREADS

| Thread Size* | Basic Major Diameter | Tap Drill Size** |
|---|---|---|
| **M1.6**x0.35 | .0630 | 1.25mm or #55 |
| **M2**x0.4 | .0787 | 1.60mm or #52 |
| **M2.5**x0.45 | .0984 | 2.05mm or #46 |
| **M3**x0.5 | .1181 | 2.50mm or #39 |
| **M3.5**x0.6 | .1378 | 2.90mm or #32 |
| **M4**x0.7 | .1575 | 3.30mm or #30 |
| **M5**x0.8 | .1969 | 4.20mm or #19 |
| **M6**x1 | .2362 | 5.00mm or #8 |
| **M8**x1.25 | .3150 | 6.80mm or H |
| M8x1 | .3150 | 7.00mm or J |
| **M10**x1.5 | .3937 | 8.50mm or R |
| M10x1.25 | .3937 | 8.80mm or 11/32 |
| **M12**x1.75 | .4724 | 10.20mm or 13/32 |
| M12x1.25 | .4724 | 10.80mm or 27/64 |
| **M14**x2 | .5512 | 12.00mm or 15/32 |
| M14x1.5 | .5512 | 12.50mm or 1/2 |
| **M16**x2 | .6299 | 14.00mm or 35/64 |
| M16x1.5 | .6299 | 14.50mm or 37/64 |
| **M18**x2.5 | .7087 | 15.50mm or 39/64 |
| M18x1.5 | .7087 | 16.50mm or 21/32 |
| **M20**x2.5 | .7874 | 17.50mm or 11/16 |
| M20x1.5 | .7874 | 18.50mm or 47/64 |
| **M22**x2.5 | .8661 | 19.50mm or 49/64 |
| M22x1.5 | .8661 | 20.50mm or 13/16 |
| **M24**x3 | .9449 | 21.00mm or 53/64 |
| M24x2 | .9449 | 22.00mm or 7/8 |
| **M27**x3 | 1.0630 | 24.00mm or 15/16 |
| M27x2 | 1.0630 | 25.00mm or 1 |

\* Coarse series shown in bold.  Pitch callout not required on metric coarse series.

\*\* Closest size for 75% theoretical thread.

# USA DRILL SIZES

| Drill | Decimal | Drill | Decimal | Drill | Decimal | Drill | Decimal |
|-------|---------|-------|---------|-------|---------|-------|---------|
| 80 | .0135 | 42 | .0935 | 13/64 | .2031 | Y | .4040 |
| 79 | .0145 | 3/32 | .0938 | 6 | .2040 | 13/32 | .4062 |
| 1/64 | .0156 | 41 | .0960 | 5 | .2055 | Z | .4130 |
| 78 | .0160 | 40 | .0980 | 4 | .2090 | 27/64 | .4219 |
| 77 | .0180 | 39 | .0995 | 3 | .2130 | **7/16** | **.4375** |
| 76 | .0200 | 38 | .1015 | 7/32 | .2188 | 29/64 | .4531 |
| 75 | .0210 | 37 | .1040 | 2 | .2210 | 15/32 | .4688 |
| 74 | .0225 | 36 | .1065 | 1 | .2280 | 31/64 | .4844 |
| 73 | .0240 | 7/64 | .1094 | A | .2340 | **1/2** | **.5000** |
| 72 | .0250 | 35 | .1100 | 15/64 | .2344 | 33/64 | .5156 |
| 71 | .0260 | 34 | .1110 | B | .2380 | 17/32 | .5312 |
| 70 | .0280 | 33 | .1130 | C | .2420 | 35/64 | .5469 |
| 69 | .0292 | 32 | .1160 | D | .2460 | **9/16** | **.5625** |
| 68 | .0310 | 31 | .1200 | E | .2500 | 37/64 | .5781 |
| 1/32 | .0312 | **1/8** | **.1250** | **1/4** | **.2500** | 19/32 | .5938 |
| 67 | .0320 | 30 | .1285 | F | .2570 | 39/64 | .6094 |
| 66 | .0330 | 29 | .1360 | G | .2610 | **5/8** | **.6250** |
| 65 | .0350 | 28 | .1405 | 17/64 | .2656 | 41/64 | .6406 |
| 64 | .0360 | 9/64 | .1406 | H | .2660 | 21/32 | .6562 |
| 63 | .0370 | 27 | .1440 | I | .2720 | 43/64 | .6719 |
| 62 | .0380 | 26 | .1470 | J | .2770 | **11/16** | **.6875** |
| 61 | .0390 | 25 | .1495 | K | .2810 | 45/64 | .7031 |
| 60 | .0400 | 24 | .1520 | 9/32 | .2812 | 23/32 | .7188 |
| 59 | .0410 | 23 | .1540 | L | .2900 | 47/64 | .7344 |
| 58 | .0420 | 5/32 | .1562 | M | .2950 | **3/4** | **.7500** |
| 57 | .0430 | 22 | .1570 | 19/64 | .2969 | 49/64 | .7656 |
| 56 | .0465 | 21 | .1590 | N | .3020 | 25/32 | .7812 |
| 3/64 | .0469 | 20 | .1610 | **5/16** | **.3125** | 51/64 | .7969 |
| 55 | .0520 | 19 | .1660 | O | .3160 | **13/16** | **.8125** |
| 54 | .0550 | 18 | .1695 | P | .3230 | 53/64 | .8281 |
| 53 | .0595 | 11/64 | .1719 | 21/64 | .3281 | 27/32 | .8438 |
| **1/16** | **.0625** | 17 | .1730 | Q | .3320 | 55/64 | .8594 |
| 52 | .0635 | 16 | .1770 | R | .3390 | **7/8** | **.8750** |
| 51 | .0670 | 15 | .1800 | 11/32 | .3438 | 57/64 | .8906 |
| 50 | .0700 | 14 | .1820 | S | .3480 | 29/32 | .9062 |
| 49 | .0730 | 13 | .1850 | T | .3580 | 59/64 | .9219 |
| 48 | .0760 | **3/16** | **.1875** | 23/64 | .3594 | **15/16** | **.9375** |
| 5/64 | .0781 | 12 | .1890 | U | .3680 | 61/64 | .9531 |
| 47 | .0785 | 11 | .1910 | **3/8** | **.3750** | 31/32 | .9688 |
| 46 | .0810 | 10 | .1935 | V | .3770 | 63/64 | .9844 |
| 45 | .0820 | 9 | .1960 | W | .3860 | **1** | **1.0000** |
| 44 | .0860 | 8 | .1990 | 25/64 | .3906 | | |
| 43 | .0890 | 7 | .2010 | X | .3970 | | |

# METRIC DRILL SIZES

| Drill | Decimal | Drill | Decimal | Drill | Decimal | Drill | Decimal |
|---|---|---|---|---|---|---|---|
| .35mm | .0138 | 2.40mm | .0945 | 5.80mm | .2283 | 9.50mm | .3740 |
| .38mm | .0150 | 2.45mm | .0965 | 5.90mm | .2323 | 9.60mm | .3780 |
| .40mm | .0157 | 2.50mm | .0984 | **6.00mm** | **.2362** | 9.70mm | .3819 |
| .42mm | .0165 | 2.55mm | .1004 | 6.10mm | .2401 | 9.75mm | .3839 |
| .45mm | .0177 | 2.60mm | .1024 | 6.20mm | .2441 | 9.80mm | .3858 |
| .48mm | .0189 | 2.65mm | .1043 | 6.25mm | .2461 | 9.90mm | .3898 |
| .50mm | .0197 | 2.70mm | .1063 | 6.30mm | .2480 | **10.00mm** | **.3937** |
| .55mm | .0217 | 2.75mm | .1083 | 6.40mm | .2520 | 10.20mm | .4016 |
| .60mm | .0236 | 2.80mm | .1102 | 6.50mm | .2559 | 10.50mm | .4134 |
| .65mm | .0256 | 2.90mm | .1142 | 6.60mm | .2598 | 10.80mm | .4252 |
| .70mm | .0276 | **3.00mm** | **.1181** | 6.70mm | .2638 | **11.00mm** | **.4330** |
| .75mm | .0295 | 3.10mm | .1220 | 6.75mm | .2658 | 11.20mm | .4409 |
| .80mm | .0315 | 3.20mm | .1260 | 6.80mm | .2677 | 11.50mm | .4528 |
| .85mm | .0335 | 3.25mm | .1280 | 6.90mm | .2716 | 11.80mm | .4646 |
| .90mm | .0354 | 3.30mm | .1299 | **7.00mm** | **.2756** | **12.00mm** | **.4724** |
| .95mm | .0374 | 3.40mm | .1339 | 7.10mm | .2795 | 12.20mm | .4803 |
| **1.00mm** | **.0394** | 3.50mm | .1378 | 7.20mm | .2835 | 12.50mm | .4921 |
| 1.05mm | .0413 | 3.60mm | .1417 | 7.25mm | .2855 | **13.00mm** | **.5118** |
| 1.10mm | .0433 | 3.70mm | .1457 | 7.30mm | .2874 | 13.50mm | .5315 |
| 1.15mm | .0453 | 3.75mm | .1477 | 7.40mm | .2913 | **14.00mm** | **.5512** |
| 1.20mm | .0472 | 3.80mm | .1496 | 7.50mm | .2953 | 14.50mm | .5709 |
| 1.25mm | .0492 | 3.90mm | .1535 | 7.60mm | .2992 | **15.00mm** | **.5906** |
| 1.30mm | .0512 | **4.00mm** | **.1575** | 7.70mm | .3031 | 15.50mm | .6102 |
| 1.35mm | .0531 | 4.10mm | .1614 | 7.75mm | .3051 | **16.00mm** | **.6299** |
| 1.40mm | .0551 | 4.20mm | .1654 | 7.80mm | .3071 | 16.50mm | .6496 |
| 1.45mm | .0571 | 4.25mm | .1674 | 7.90mm | .3110 | **17.00mm** | **.6693** |
| 1.50mm | .0591 | 4.30mm | .1693 | **8.00mm** | **.3150** | 17.50mm | .6890 |
| 1.55mm | .0610 | 4.40mm | .1732 | 8.10mm | .3189 | **18.00mm** | **.7087** |
| 1.60mm | .0629 | 4.50mm | .1771 | 8.20mm | .3228 | 18.50mm | .7283 |
| 1.65mm | .0650 | 4.60mm | .1811 | 8.25mm | .3248 | **19.00mm** | **.7480** |
| 1.70mm | .0669 | 4.70mm | .1850 | 8.30mm | .3268 | 19.50mm | .7677 |
| 1.75mm | .0689 | 4.75mm | .1870 | 8.40mm | .3307 | **20.00mm** | **.7874** |
| 1.80mm | .0709 | 4.80mm | .1890 | 8.50mm | .3346 | 20.50mm | .8071 |
| 1.85mm | .0728 | 4.90mm | .1929 | 8.60mm | .3386 | **21.00mm** | **.8268** |
| 1.90mm | .0748 | **5.00mm** | **.1968** | 8.70mm | .3425 | 21.50mm | .8465 |
| 1.95mm | .0768 | 5.10mm | .2008 | 8.75mm | .3445 | **22.00mm** | **.8661** |
| **2.00mm** | **.0787** | 5.20mm | .2047 | 8.80mm | .3465 | 22.50mm | .8858 |
| 2.05mm | .0807 | 5.25mm | .2067 | 8.90mm | .3504 | **23.00mm** | **.9055** |
| 2.10mm | .0827 | 5.30mm | .2087 | **9.00mm** | **.3543** | 23.50mm | .9252 |
| 2.15mm | .0846 | 5.40mm | .2126 | 9.10mm | .3583 | **24.00mm** | **.9449** |
| 2.20mm | .0866 | 5.50mm | .2165 | 9.20mm | .3622 | 24.50mm | .9646 |
| 2.25mm | .0886 | 5.60mm | .2205 | 9.25mm | .3642 | **25.00mm** | **.9843** |
| 2.30mm | .0905 | 5.70mm | .2244 | 9.30mm | .3661 | | |
| 2.35mm | .0925 | 5.75mm | .2264 | 9.40mm | .3701 | | |

# COUNTERBORED HOLES

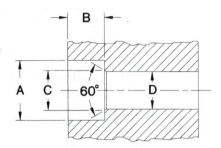

## USA SOCKET-HEAD CAP SCREWS

| SCREW DIA | A COUNTERBORE DIA | B COUNTERBORE DEPTH | C COUNTERSINK DIA | D CLEARANCE DIA | |
|---|---|---|---|---|---|
| | | | | NORMAL FIT | CLOSE FIT |
| #0 | 1/8 | .060 | .074 | #49 | #51 |
| #2 | 3/16 | .086 | .102 | #36 | 3/32 |
| #4 | 7/32 | .112 | .130 | #29 | 1/8 |
| #5 | 1/4 | .125 | .145 | #23 | 9/64 |
| #6 | 9/32 | .138 | .158 | #18 | #23 |
| #8 | 5/16 | .164 | .188 | #10 | #15 |
| #10 | 3/8 | .190 | .218 | #2 | #5 |
| 1/4 | 7/16 | .250 | .278 | 9/32 | 17/64 |
| 5/16 | 17/32 | .312 | .346 | 11/32 | 21/64 |
| 3/8 | 5/8 | .375 | .415 | 13/32 | 25/64 |
| 7/16 | 23/32 | .438 | .483 | 15/32 | 29/64 |
| 1/2 | 13/16 | .500 | .552 | 17/32 | 33/64 |
| 5/8 | 1 | .625 | .689 | 21/32 | 41/64 |
| 3/4 | 1-3/16 | .750 | .828 | 25/32 | 49/64 |
| 7/8 | 1-3/8 | .875 | .963 | 29/32 | 57/64 |
| 1 | 1-5/8 | 1.000 | 1.100 | 1-1/32 | 1-1/64 |
| 1-1/4 | 2 | 1.250 | 1.370 | 1-5/16 | 1-9/32 |
| 1-1/2 | 2-3/8 | 1.500 | 1.640 | 1-9/16 | 1-17/32 |
| 1-3/4 | 2-3/4 | 1.750 | 1.910 | 1-13/16 | 1-25/32 |
| 2 | 3-1/8 | 2.000 | 2.180 | 2-1/16 | 2-1/32 |

## METRIC SOCKET-HEAD CAP SCREWS

| | | | | | |
|---|---|---|---|---|---|
| M1.6 | 3.50mm | 1.6mm | 2.0mm | 1.95mm | 1.80mm |
| M2 | 4.40mm | 2mm | 2.6mm | 2.40mm | 2.20mm |
| M2.5 | 5.40mm | 2.5mm | 3.1mm | 3.00mm | 2.70mm |
| M3 | 6.50mm | 3mm | 3.6mm | 3.70mm | 3.40mm |
| M4 | 8.25mm | 4mm | 4.7mm | 4.80mm | 4.40mm |
| M5 | 9.75mm | 5mm | 5.7mm | 5.80mm | 5.40mm |
| M6 | 11.20mm | 6mm | 6.8mm | 6.80mm | 6.40mm |
| M8 | 14.50mm | 8mm | 9.2mm | 8.80mm | 8.40mm |
| M10 | 17.50mm | 10mm | 11.2mm | 10.80mm | 10.50mm |
| M12 | 19.50mm | 12mm | 14.2mm | 13.00mm | 12.50mm |
| M14 | 22.50mm | 14mm | 16.2mm | 15.00mm | 14.50mm |
| M16 | 25.50mm | 16mm | 18.2mm | 17.00mm | 16.50mm |
| M20 | 31.50mm | 20mm | 22.4mm | 21.00mm | 20.50mm |
| M24 | 37.50mm | 24mm | 26.4mm | 25.00mm | 24.50mm |
| M30 | 47.50mm | 30mm | 33.4mm | 31.50mm | 31.00mm |
| M36 | 56.50mm | 36mm | 39.4mm | 37.50mm | 37.00mm |
| M42 | 66.00mm | 42mm | 45.6mm | 44.00mm | 43.00mm |
| M48 | 75.00mm | 48mm | 52.6mm | 50.00mm | 49.00mm |

# GEOMETRIC SYMBOLS AND DEFINITIONS

Individual Features

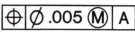

Related Features

## Feature-Control Frame

A specification box that shows a particular geometric characteristic (flatness, straightness, etc.) applied to a part feature and states the allowable tolerance. The feature's tolerance may be individual, or related to one or more datums. Any datum references and tolerance modifiers are also shown.

### Datum Feature

A flag which designates a physical feature of the part to be used as a reference to measure geometric characteristics of other part features.

### Datum Targets

Callouts occasionally needed to designate specific points, lines, or areas on an actual part to be used to establish a theoretical datum feature.

### Basic Dimension

A box around any drawing dimension makes it a "basic" dimension, a theoretically exact value used as a reference for measuring geometric characteristics and tolerances of other part features.

### Cylindrical Tolerance Zone

This symbol, commonly used to indicate a diameter dimension, also specifies a cylindrically shaped tolerance zone in a feature-control frame.

### Maximum Material Condition

Abbreviation: MMC. A tolerance modifier that applies the stated tight tolerance zone only while the part theoretically contains the maximum amount of material permitted within its dimensional limits (e.g. minimum hole diameters and maximum shaft diameters), allowing more variation under normal conditions.

### Least Material Condition

Abbreviation: LMC. A tolerance modifier that applies the stated tight tolerance zone only while the part theoretically contains the minimum amount of material permitted within its dimensional limits (e.g. maximum hole diameters and minimum shaft diameters), allowing more variation under normal conditions.

### Regardless of Feature Size

Abbreviation: RFS. A tolerance modifier that applies the stated tight tolerance zone under all size conditions. RFS is generally assumed if neither MMC nor LMC are stated.

### Projected Tolerance Zone

An additional specification box attached underneath a feature-control frame. It extends the feature's tolerance zone beyond the part's surface by the stated distance, ensuring perpendicularity for proper alignment of mating parts.

# GEOMETRIC CHARACTERISTICS

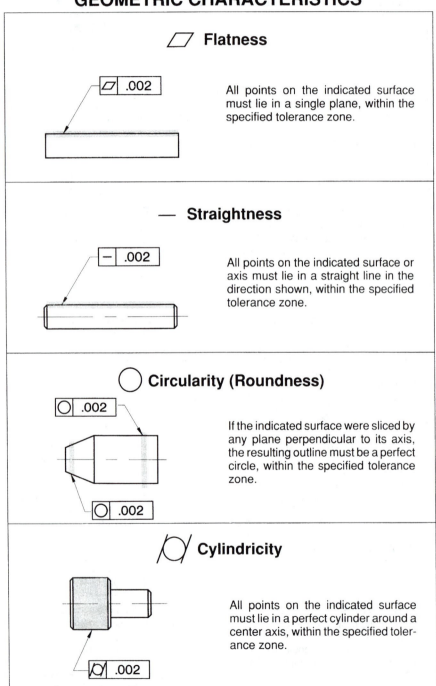

## ▱ Flatness

⬜ .002

All points on the indicated surface must lie in a single plane, within the specified tolerance zone.

## — Straightness

— .002

All points on the indicated surface or axis must lie in a straight line in the direction shown, within the specified tolerance zone.

## ◯ Circularity (Roundness)

◯ .002

◯ .002

If the indicated surface were sliced by any plane perpendicular to its axis, the resulting outline must be a perfect circle, within the specified tolerance zone.

## ⌭ Cylindricity

⌭ .002

All points on the indicated surface must lie in a perfect cylinder around a center axis, within the specified tolerance zone.

# GEOMETRIC CHARACTERISTICS

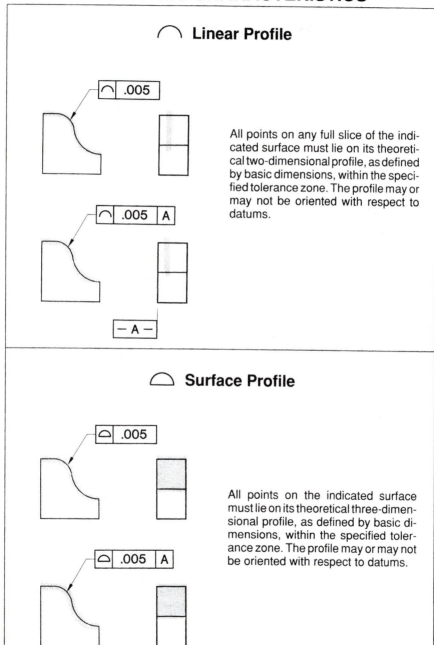

## ⌒ Linear Profile

⌒ .005

⌒ .005 | A

− A −

All points on any full slice of the indicated surface must lie on its theoretical two-dimensional profile, as defined by basic dimensions, within the specified tolerance zone. The profile may or may not be oriented with respect to datums.

## ◠ Surface Profile

◠ .005

◠ .005 | A

− A −

All points on the indicated surface must lie on its theoretical three-dimensional profile, as defined by basic dimensions, within the specified tolerance zone. The profile may or may not be oriented with respect to datums.

# GEOMETRIC CHARACTERISTICS

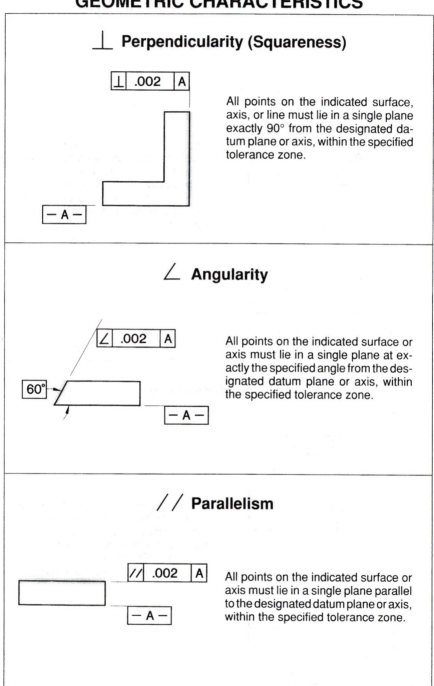

## ⊥ Perpendicularity (Squareness)

| ⊥ | .002 | A |

All points on the indicated surface, axis, or line must lie in a single plane exactly 90° from the designated datum plane or axis, within the specified tolerance zone.

— A —

## ∠ Angularity

| ∠ | .002 | A |

All points on the indicated surface or axis must lie in a single plane at exactly the specified angle from the designated datum plane or axis, within the specified tolerance zone.

60°

— A —

## // Parallelism

| // | .002 | A |

All points on the indicated surface or axis must lie in a single plane parallel to the designated datum plane or axis, within the specified tolerance zone.

— A —

# GEOMETRIC CHARACTERISTICS

## ↗ Circular Runout

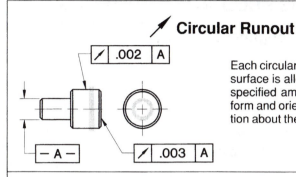

Each circular element of the indicated surface is allowed to deviate only the specified amount from its theoretical form and orientation during 360° rotation about the designated datum axis.

## ↗↗ Total Runout

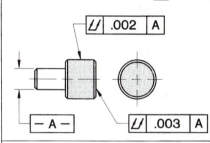

The entire indicated surface is allowed to deviate only the specified amount from its theoretical form and orientation during 360° rotation about the designated datum axis.

## ◎ Concentricity

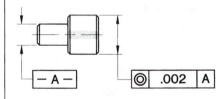

If the indicated surface were sliced by any plane perpendicular to the designated datum axis, every slice's center of area must lie on the datum axis, within the specified cylindrical tolerance zone (controls rotational balance).

## ⊕ Position (Replaces ═ Symmetry)

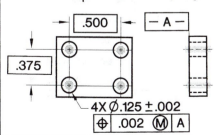

The indicated feature's axis must be located within the specified tolerance zone from its true theoretical position, correctly oriented relative to the designated datum plane or axis.

# TYPICAL SURFACE FINISHES PRODUCED BY VARIOUS PRODUCTION METHODS

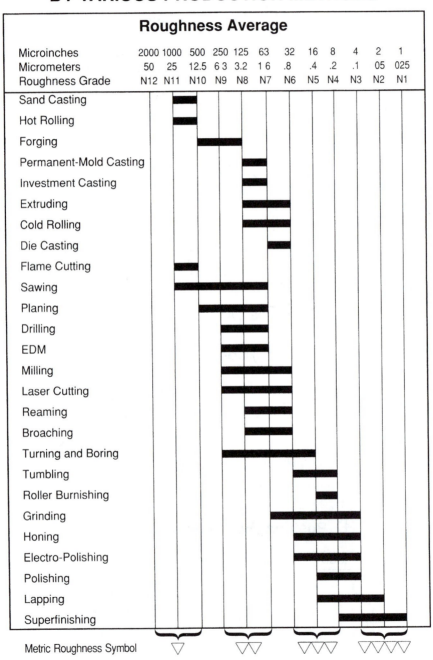

## Roughness Average

| | | | | | | | | | | | | |
|---|---|---|---|---|---|---|---|---|---|---|---|---|
| Microinches | 2000 | 1000 | 500 | 250 | 125 | 63 | 32 | 16 | 8 | 4 | 2 | 1 |
| Micrometers | 50 | 25 | 12.5 | 6 3 | 3.2 | 1 6 | .8 | .4 | .2 | .1 | 05 | 025 |
| Roughness Grade | N12 | N11 | N10 | N9 | N8 | N7 | N6 | N5 | N4 | N3 | N2 | N1 |

Sand Casting

Hot Rolling

Forging

Permanent-Mold Casting

Investment Casting

Extruding

Cold Rolling

Die Casting

Flame Cutting

Sawing

Planing

Drilling

EDM

Milling

Laser Cutting

Reaming

Broaching

Turning and Boring

Tumbling

Roller Burnishing

Grinding

Honing

Electro-Polishing

Polishing

Lapping

Superfinishing

Metric Roughness Symbol

# CLAMPING FORCES
# OF STANDARD CLAMP STRAPS

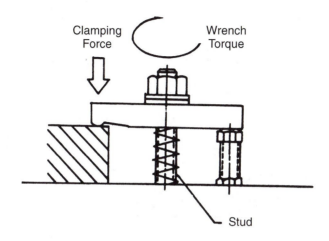

## USA

| Stud Size | Recommended Torque* (ft-lbs) | Clamping Force (lbs) | Tensile Force in Stud (lbs) |
|---|---|---|---|
| #10-32 | 2 | 300 | 600 |
| 1/4-20 | 4 | 500 | 1000 |
| 5/16-18 | 9 | 900 | 1800 |
| 3/8-16 | 16 | 1300 | 2600 |
| 1/2-13 | 38 | 2300 | 4600 |
| 5/8-11 | 77 | 3700 | 7400 |
| 3/4-10 | 138 | 5500 | 11000 |
| 7/8-9 | 222 | 7600 | 15200 |
| 1-8 | 333 | 10000 | 20000 |

## METRIC

| | | | |
|---|---|---|---|
| M6 | 4 | 500 | 1000 |
| M8 | 9 | 900 | 1800 |
| M10 | 20 | 1500 | 3000 |
| M12 | 35 | 2200 | 4400 |
| M16 | 84 | 4000 | 8000 |
| M20 | 165 | 6300 | 12600 |
| M24 | 283 | 9000 | 18000 |

* Clean, dry clamping stud torqued to approximately 33% of its 100,000 psi yield strength (2:1 lever ratio).

# AISI/SAE NUMBERING SYSTEM
# FOR CARBON AND ALLOY STEELS

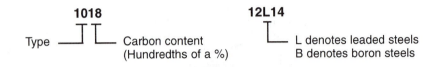

| | | |
|---|---|---|
| | 10XX | Plain Carbon, Mn 1.00% max |
| | 11XX | Resulfurized Free Machining |
| Carbon Steels | 12XX | Resulfurized/Rephosphorized Free Machining |
| | 15XX | Plain Carbon, Mn 1.00-1.65% |
| Manganese Steels | 13XX | Mn 1.75% |
| Nickel Steels | 23XX | Ni 3.50% |
| | 25XX | Ni 5.00% |
| | 31XX | Ni 1.25%, Cr .65-.80% |
| Nickel-Chromium Steels | 32XX | Ni 1.75%, Cr 1.07% |
| | 33XX | Ni 3.50%, Cr 1.50-1.57% |
| | 34XX | Ni 3.00%, Cr .77% |
| Molybdenum Steels | 40XX | Mo .20-.25% |
| | 44XX | Mo .40-.52% |
| Chromium-Molybdenum Steels | 41XX | Cr .50-.95%, Mo .12-.30% |
| Nickel-Chromium-Molybdenum Steels | 43XX | Ni 1.82%, Cr .50-.80%, Mo .25% |
| | 47XX | Ni 1.05%, Cr .45%, Mo .20-.35% |
| Nickel-Molybdenum Steels | 46XX | Ni .85-1.82%, Mo .20-.25% |
| | 48XX | Ni 3.50%, Mo .25% |
| | 50XX | Cr .27-.65% |
| | 51XX | Cr .80-1.05% |
| Chromium Steels | 50XXX | Cr .50%, C 1.00% min |
| | 51XXX | Cr 1.02%, C 1.00% min |
| | 52XXX | Cr 1.45%, C 1.00% min |
| Chromium-Vanadium Steels | 61XX | Cr .60-.95%, V .10-.15% |
| Tungsten-Chromium Steels | 72XX | W 1.75%, Cr .75% |
| | 81XX | Ni .30%, Cr .40%, Mo .12% |
| Nickel-Chromium-Molybdenum Steels | 86XX | Ni .55%, Cr .50%, Mo .20% |
| | 87XX | Ni .55%, Cr .50%, Mo .25% |
| | 88XX | Ni .55%, Cr .50%, Mo .35% |
| Silicon-Manganese Steels | 92XX | Si 1.40-2.00%, Mn .65-.85%, Cr 0-.65% |
| | 93XX | Ni 3.25%, Cr 1.20%, Mo .12% |
| Nickel-Chromium-Molybdenum Steels | 94XX | Ni .45%, Cr .40%, Mo .12% |
| | 97XX | Ni .55%, Cr .20%, Mo .20% |
| | 98XX | Ni 1.00%, Cr .80%, Mo .25% |

# MACHINABILITY COMPARISON*

**Carbon Steels:**

| | | | |
|---|---|---|---|
| 1015 | 72% | 1137 | 72% |
| 1018 | 78% | 1141 | 70% |
| 1020 | 72% | 1141 Annealed | 81% |
| 1022 | 78% | 1144 | 76% |
| 1030 | 70% | 1144 Annealed | 85% |
| 1040 | 64% | 1144 Stressproof | 83% |
| 1042 | 64% | 1212 | 100% |
| 1050 | 54% | 1213 | 136% |
| 1095 | 42% | 12L14 | 170% |
| 1117 | 91% | 1215 | 136% |

**Alloy Steels:**

| | | | |
|---|---|---|---|
| 2355 Annealed | 70% | 4620 | 66% |
| 4130 Annealed | 72% | 4820 Annealed | 49% |
| 4140 Annealed | 66% | 52100 Annealed | 40% |
| 4142 Annealed | 66% | 6150 Annealed | 60% |
| 41L42 Annealed | 77% | 8620 | 66% |
| 4150 Annealed | 60% | 86L20 | 77% |
| 4340 Annealed | 57% | 9310 Annealed | 51% |

**Stainless Steels and Super Alloys:**

| | | | |
|---|---|---|---|
| 302 Annealed | 45% | 420 Annealed | 45% |
| 303 Annealed | 78% | 430 Annealed | 54% |
| 304 Annealed | 45% | 431 Annealed | 45% |
| 316 Annealed | 45% | 440A | 45% |
| 321 Annealed | 36% | 15-5PH Condition A | 48% |
| 347 Annealed | 36% | 17-4PH Condition A | 48% |
| 410 Annealed | 54% | A286 Aged | 33% |
| 416 Annealed | 110% | Hastelloy X | 19% |

**Tool Steels:**

| | | | |
|---|---|---|---|
| A-2 | 42% | M-2 | 39% |
| A-6 | 33% | O-1 | 42% |
| D-2 | 27% | O-2 | 42% |
| D-3 | 27% | | |

**Gray Cast Iron:**

| | | | |
|---|---|---|---|
| ASTM Class 20 Annealed | 73% | ASTM Class 40 | 48% |
| ASTM Class 25 | 55% | ASTM Class 45 | 36% |
| ASTM Class 30 | 48% | ASTM Class 50 | 36% |
| ASTM Class 35 | 48% | | |

**Nodular Iron:**

| | | | |
|---|---|---|---|
| 60-40-18 Annealed | 61% | 80-55-06 | 39% |
| 65-45-12 Annealed | 61% | | |

**Aluminum and Magnesium Alloys:**

| | | | |
|---|---|---|---|
| Aluminum, Cold Drawn | 360% | Magnesium, Cold Drawn | 480% |
| Aluminum, Cast | 450% | Magnesium, Cast | 480% |
| Aluminum, Die Cast | 76% | | |

*Relative machining speed based on 1212 as 100%. All figures are based on cold-drawn bars in as-drawn condition, except where noted.

# APPROXIMATE HARDNESS CONVERSIONS FOR STEEL

| Rockwell | | | | | | Rockwell Superficial | | | | Brinell | | Vickers | Shore | Approximate Tensile Strength (psi) |
|---|---|---|---|---|---|---|---|---|---|---|---|---|---|---|
| A 60 kg Brale | B 150 kg 1/16" Ball | C 150 kg Brale | D 100 kg Brale | E 100 kg 1/8" Ball | F 60 kg 1/16" Ball | 15-N 15 kg Brale | 30-N 30 kg Brale | 45-N 45 kg Brale | 30-T 30 kg 1/16" Ball | 3000 kg 10mm Ball Steel* | 500 kg 10mm Ball Steel | 136° Diamond Pyramid | Sclero-scope | |
| 86.5 | — | 70 | 78.5 | — | — | 94.0 | 86.0 | 77.6 | — | — | — | 1076 | 101 | — |
| 86.0 | — | 69 | 77.7 | — | — | 93.5 | 85.0 | 76.5 | — | — | — | 1044 | 99 | — |
| 85.6 | — | 68 | 76.9 | — | — | 93.2 | 84.4 | 75.4 | — | — | — | 940 | 97 | — |
| 85.0 | — | 67 | 76.1 | — | — | 92.9 | 83.6 | 74.2 | — | — | — | 900 | 95 | — |
| 84.5 | — | 66 | 75.4 | — | — | 92.5 | 82.8 | 73.2 | — | — | — | 865 | 92 | — |
| 83.9 | — | 65 | 74.5 | — | — | 92.2 | 81.9 | 72.0 | — | 739 | — | 832 | 91 | — |
| 83.4 | — | 64 | 73.8 | — | — | 91.8 | 81.1 | 71.0 | — | 722 | — | 800 | 88 | — |
| 82.8 | — | 63 | 73.0 | — | — | 91.4 | 80.1 | 69.9 | — | 705 | — | 772 | 87 | — |
| 82.3 | — | 62 | 72.2 | — | — | 91.1 | 79.3 | 68.8 | — | 688 | — | 746 | 85 | — |
| 81.8 | — | 61 | 71.5 | — | — | 90.7 | 78.4 | 67.7 | — | 670 | — | 720 | 83 | — |
| 81.2 | — | 60 | 70.7 | — | — | 90.2 | 77.5 | 66.6 | — | 654 | — | 697 | 81 | 320,000 |
| 80.7 | — | 59 | 69.9 | — | — | 89.8 | 76.6 | 65.5 | — | 634 | — | 674 | 80 | 310,000 |
| 80.1 | — | 58 | 69.2 | — | — | 89.3 | 75.7 | 64.3 | — | 615 | — | 653 | 78 | 300,000 |
| 79.6 | — | 57 | 68.5 | — | — | 88.9 | 74.8 | 63.2 | — | 595 | — | 633 | 76 | 290,000 |
| 79.0 | — | 56 | 67.7 | — | — | 88.3 | 73.9 | 62.0 | — | 577 | — | 613 | 75 | 282,000 |
| 78.5 | 120 | 55 | 66.9 | — | — | 87.9 | 73.0 | 60.9 | — | 560 | — | 595 | 74 | 274,000 |
| 78.0 | 120 | 54 | 66.1 | — | — | 87.4 | 72.0 | 59.8 | — | 543 | — | 577 | 72 | 266,000 |
| 77.4 | 119 | 53 | 65.4 | — | — | 86.9 | 71.2 | 58.6 | — | 525 | — | 560 | 71 | 257,000 |
| 76.8 | 119 | 52 | 64.6 | — | — | 86.4 | 70.2 | 57.4 | — | 500 | — | 544 | 69 | 245,000 |
| 76.3 | 118 | 51 | 63.8 | — | — | 85.9 | 69.4 | 56.1 | — | 487 | — | 528 | 68 | 239,000 |
| 75.9 | 117 | 50 | 63.1 | — | — | 85.5 | 68.5 | 55.0 | — | 475 | — | 513 | 67 | 233,000 |
| 75.2 | 117 | 49 | 62.1 | — | — | 85.0 | 67.6 | 53.8 | — | 464 | — | 498 | 66 | 227,000 |
| 74.7 | 116 | 48 | 61.4 | — | — | 84.5 | 66.7 | 52.5 | — | 451 | — | 484 | 64 | 221,000 |
| 74.1 | 116 | 47 | 60.8 | — | — | 83.9 | 65.8 | 51.4 | — | 442 | — | 471 | 63 | 217,000 |
| 73.6 | 115 | 46 | 60.0 | — | — | 83.5 | 64.8 | 50.3 | — | 432 | — | 458 | 62 | 212,000 |

*Brinell hardness values above 500 are listed for a 10mm carbide ball instead of a steel ball.

## APPROXIMATE HARDNESS CONVERSIONS FOR STEEL

| Rockwell | | | | | | Rockwell Superficial | | | | Brinell | | Vickers | Shore | Approximate Tensile Strength |
|---|---|---|---|---|---|---|---|---|---|---|---|---|---|---|
| A 60 kg Brale | B 150 kg 1/16" Ball | C 150 kg Brale | D 100 kg Brale | E 100 kg 1/8" Ball | F 60 kg 1/16" Ball | 15-N 15 kg Brale | 30-N 30 kg Brale | 45-N 45 kg Brale | 30-T 30 kg 1/16" Ball | 3000 kg 10mm Ball Steel | 500 kg 10mm Ball Steel | 136° Diamond Pyramid | Sclero-scope | (psi) |
| 73.1 | 115 | 45 | 59.2 | — | — | 83.0 | 64.0 | 49.0 | — | 421 | — | 446 | 60 | 206,000 |
| 72.5 | 114 | 44 | 58.5 | — | — | 82.5 | 63.1 | 47.8 | — | 409 | — | 434 | 58 | 200,000 |
| 72.0 | 113 | 43 | 57.7 | — | — | 82.0 | 62.2 | 46.7 | — | 400 | — | 423 | 57 | 196,000 |
| 71.5 | 113 | 42 | 56.9 | — | — | 81.5 | 61.3 | 45.5 | — | 390 | — | 412 | 56 | 191,000 |
| 70.9 | 112 | 41 | 56.2 | — | — | 80.9 | 60.4 | 44.3 | — | 381 | — | 402 | 55 | 187,000 |
| 70.4 | 112 | 40 | 55.4 | — | — | 80.4 | 59.5 | 43.1 | — | 371 | — | 392 | 54 | 182,000 |
| 69.9 | 111 | 39 | 54.6 | — | — | 79.9 | 58.6 | 41.9 | — | 362 | — | 382 | 52 | 177,000 |
| 69.4 | 110 | 38 | 53.8 | — | — | 79.4 | 57.7 | 40.8 | — | 353 | — | 372 | 51 | 173,000 |
| 68.9 | 110 | 37 | 53.1 | — | — | 78.8 | 56.8 | 39.6 | — | 344 | — | 363 | 50 | 169,000 |
| 68.4 | 109 | 36 | 52.3 | — | — | 78.3 | 55.9 | 38.4 | — | 336 | — | 354 | 49 | 165,000 |
| 67.9 | 109 | 35 | 51.5 | — | — | 77.7 | 55.0 | 37.2 | — | 327 | — | 345 | 48 | 160,000 |
| 67.4 | 108 | 34 | 50.8 | — | — | 77.2 | 54.2 | 36.1 | — | 319 | — | 336 | 47 | 156,000 |
| 66.8 | 108 | 33 | 50.0 | — | — | 76.6 | 53.3 | 34.9 | — | 311 | — | 327 | 46 | 152,000 |
| 66.3 | 107 | 32 | 49.2 | — | — | 76.1 | 52.1 | 33.7 | — | 301 | — | 318 | 44 | 147,000 |
| 65.8 | 106 | 31 | 48.4 | — | — | 75.6 | 51.3 | 32.5 | — | 294 | — | 310 | 43 | 144,000 |
| 65.3 | 105 | 30 | 47.7 | — | — | 75.0 | 50.4 | 31.3 | — | 286 | — | 302 | 42 | 140,000 |
| 64.7 | 104 | 29 | 47.0 | — | — | 74.5 | 49.5 | 30.1 | — | 279 | — | 294 | 41 | 137,000 |
| 64.3 | 104 | 28 | 46.1 | — | — | 73.9 | 48.6 | 28.9 | — | 271 | — | 286 | 41 | 133,000 |
| 63.8 | 103 | 27 | 45.2 | — | — | 73.3 | 47.7 | 27.8 | — | 264 | — | 279 | 40 | 129,000 |
| 63.3 | 103 | 26 | 44.6 | — | — | 72.8 | 46.8 | 26.7 | — | 258 | — | 272 | 39 | 126,000 |
| 62.8 | 102 | 25 | 43.8 | — | — | 72.2 | 45.9 | 25.5 | — | 253 | — | 266 | 38 | 124,000 |
| 62.4 | 101 | 24 | 43.1 | — | — | 71.6 | 45.0 | 24.3 | — | 247 | — | 260 | 37 | 121,000 |
| 62.0 | **100** | 23 | 42.1 | — | — | 71.0 | 44.0 | 23.1 | 82.0 | 240 | 201 | 254 | 36 | 118,000 |
| 61.5 | 99 | 22 | 41.6 | — | — | 70.5 | 43.2 | 22.0 | 81.5 | 234 | 195 | 248 | 35 | 115,000 |
| 61.0 | 98 | 21 | 40.9 | — | — | 69.9 | 42.3 | 20.7 | 81.0 | 228 | 189 | 243 | 35 | 112,000 |

## APPROXIMATE HARDNESS CONVERSIONS FOR STEEL

| A 60 kg Brale | B 150 kg 1/16" Ball | C 150 kg Brale | D 100 kg Brale | E 100 kg 1/8" Ball | F 60 kg 1/16" Ball | 15-N 15 kg Brale | 30-N 30 kg Brale | 45-N 45 kg Brale | 30-T 30 kg 1/16" Ball | Brinell 3000 kg 10mm Ball Steel | Brinell 500 kg 10mm Ball Steel | Vickers 136° Diamond Pyramid | Shore Scleroscope | Approximate Tensile Strength (psi) |
|---|---|---|---|---|---|---|---|---|---|---|---|---|---|---|
| 60.5 | 97 | 20 | 40.1 | — | — | 69.4 | 41.5 | 19.6 | 80.5 | 222 | 184 | 238 | 34 | 109,000 |
| 59.0 | 96 | 18 | — | — | — | — | — | — | 80.0 | 216 | 179 | 230 | 33 | 106,000 |
| 58.0 | 95 | 16 | — | — | — | — | — | — | 79.0 | 210 | 175 | 222 | 32 | 103,000 |
| 57.5 | 94 | 15 | — | — | — | — | — | — | 78.5 | 205 | 171 | 213 | 31 | 100,000 |
| 57.0 | 93 | 13 | — | — | — | — | — | — | 78.0 | 200 | 167 | 208 | 30 | 98,000 |
| 56.5 | 92 | 12 | — | — | — | — | — | — | 77.5 | 195 | 163 | 204 | 29 | 96,000 |
| 56.0 | 91 | 10 | — | — | — | — | — | — | 77.0 | 190 | 160 | 196 | 28 | 93,000 |
| 55.5 | 90 | 9 | — | — | — | — | — | — | 76.0 | 185 | 157 | 192 | 27 | 91,000 |
| 55.0 | 89 | 8 | — | — | — | — | — | — | 75.5 | 180 | 154 | 188 | 26 | 88,000 |
| 54.0 | 88 | 7 | — | — | — | — | — | — | 75.0 | 176 | 151 | 184 | 26 | 86,000 |
| 53.5 | 87 | 6 | — | — | — | — | — | — | 74.5 | 172 | 148 | 180 | 26 | 84,000 |
| 53.0 | 86 | 5 | — | — | — | — | — | — | 74.0 | 169 | 145 | 176 | 25 | 83,000 |
| 52.5 | 85 | 4 | — | — | — | — | — | — | 73.5 | 165 | 142 | 173 | 25 | 81,000 |
| 52.0 | 84 | 3 | — | — | — | — | — | — | 73.0 | 162 | 140 | 170 | 25 | 79,000 |
| 51.0 | 83 | 2 | — | — | — | — | — | — | 72.0 | 159 | 137 | 166 | 24 | 78,000 |
| 50.5 | 82 | 1 | — | — | — | — | — | — | 71.5 | 156 | 135 | 163 | 24 | 76,000 |
| 50.0 | 81 | 0 | — | — | — | — | — | — | 71.0 | 153 | 133 | 160 | 24 | 75,000 |
| 49.5 | 80 | — | — | — | — | — | — | — | 70.0 | 150 | 130 | — | — | 73,000 |
| 49.0 | 79 | — | — | — | — | — | — | — | 69.5 | 147 | 128 | — | — | — |
| 48.5 | 78 | — | — | — | — | — | — | — | 69.0 | 144 | 126 | — | — | — |
| 48.0 | 77 | — | — | — | — | — | — | — | 68.0 | 141 | 124 | — | — | — |
| 47.0 | 76 | — | — | — | — | — | — | — | 67.5 | 139 | 122 | — | — | — |
| 46.5 | 75 | — | — | — | 99.5 | — | — | — | 67.0 | 137 | 120 | — | — | — |
| 46.0 | 74 | — | — | — | 99.0 | — | — | — | 66.0 | 135 | 118 | — | — | — |
| 45.5 | 73 | — | — | — | 98.5 | — | — | — | 65.5 | 132 | 116 | — | — | — |

## APPROXIMATE HARDNESS CONVERSIONS FOR STEEL

| Rockwell | | | | | | Rockwell Superficial | | | | Brinell | | Vickers | Shore | Approximate Tensile Strength (psi) |
|---|---|---|---|---|---|---|---|---|---|---|---|---|---|---|
| A 60 kg Brale | B 150 kg 1/16" Ball | C 150 kg Brale | D 100 kg Brale | E 100 kg 1/8" Ball | F 60 kg 1/16" Ball | 15-N 15 kg Brale | 30-N 30 kg Brale | 45-N 45 kg Brale | 30-T 30 kg 1/16" Ball | 3000 kg 10mm Ball Steel | 500 kg 10mm Ball Steel | 136° Diamond Pyramid | Sclero-scope | |
| 45.0 | 72 | — | — | — | 98.0 | — | — | — | 65.0 | 130 | 114 | — | — | — |
| 44.5 | 71 | — | — | 100.0 | 97.5 | — | — | — | 64.2 | 127 | 112 | — | — | — |
| 44.0 | 70 | — | — | 99.5 | 97.0 | — | — | — | 63.5 | 125 | 110 | — | — | — |
| 43.5 | 69 | — | — | 99.0 | 96.0 | — | — | — | 62.8 | 123 | 109 | — | — | — |
| 43.0 | 68 | — | — | 98.0 | 95.5 | — | — | — | 62.0 | 121 | 107 | — | — | — |
| 42.5 | 67 | — | — | 97.5 | 95.0 | — | — | — | 61.4 | 119 | 106 | — | — | — |
| 42.0 | 66 | — | — | 97.0 | 94.5 | — | — | — | 60.5 | 117 | 104 | — | — | — |
| 41.8 | 65 | — | — | 96.0 | 94.0 | — | — | — | 60.1 | 116 | 102 | — | — | — |
| 41.5 | 64 | — | — | 95.5 | 93.5 | — | — | — | 59.5 | 114 | 101 | — | — | — |
| 41.0 | 63 | — | — | 95.0 | 93.0 | — | — | — | 58.7 | 112 | 99 | — | — | — |
| 40.5 | 62 | — | — | 94.5 | 92.0 | — | — | — | 58.0 | 110 | 98 | — | — | — |
| 40.0 | 61 | — | — | 93.5 | 91.5 | — | — | — | 57.3 | 108 | 96 | — | — | — |
| 39.5 | 60 | — | — | 93.0 | 91.0 | — | — | — | 56.5 | 107 | 95 | — | — | — |
| 39.0 | 59 | — | — | 92.5 | 90.5 | — | — | — | 55.9 | 106 | 94 | — | — | — |
| 38.5 | 58 | — | — | 92.0 | 90.0 | — | — | — | 55.0 | 104 | 92 | — | — | — |
| 38.0 | 57 | — | — | 91.0 | 89.5 | — | — | — | 54.6 | 102 | 91 | — | — | — |
| 37.8 | 56 | — | — | 90.5 | 89.0 | — | — | — | 54.0 | 101 | 90 | — | — | — |
| 37.5 | 55 | — | — | 90.0 | 88.0 | — | — | — | 53.2 | 99 | 89 | — | — | — |
| 37.0 | 54 | — | — | 89.5 | 87.5 | — | — | — | 52.5 | — | 87 | — | — | — |
| 36.5 | 53 | — | — | 89.0 | 87.0 | — | — | — | 51.8 | — | 86 | — | — | — |
| 36.0 | 52 | — | — | 88.0 | 86.5 | — | — | — | 51.0 | — | 85 | — | — | — |
| 35.5 | 51 | — | — | 87.5 | 86.0 | — | — | — | 50.4 | — | 84 | — | — | — |
| 35.0 | 50 | — | — | 87.0 | 85.5 | — | — | — | 49.5 | — | 83 | — | — | — |
| 34.8 | 49 | — | — | 86.5 | 85.0 | — | — | — | 49.1 | — | 82 | — | — | — |
| 34.5 | 48 | — | — | 85.5 | 84.5 | — | — | — | 48.5 | — | 81 | — | — | — |

## APPROXIMATE HARDNESS CONVERSIONS FOR STEEL

| Rockwell | | | | | | Rockwell Superficial | | | | Brinell | | Vickers | Shore | Approximate Tensile Strength (psi) |
| A 60 kg Brale | B 150 kg 1/16" Ball | C 150 kg Brale | D 100 kg Brale | E 100 kg 1/8" Ball | F 60 kg 1/16" Ball | 15-N 15 kg Brale | 30-N 30 kg Brale | 45-N 45 kg Brale | 30-T 30 kg 1/16" Ball | 3000 kg 10mm Ball Steel | 500 kg 10mm Ball Steel | 136° Diamond Pyramid | Sclero-scope | |
|---|---|---|---|---|---|---|---|---|---|---|---|---|---|---|
| 34.0 | 47 | — | — | 85.0 | 84.0 | — | — | — | 47.7 | — | 80 | — | — | — |
| 33.5 | 46 | — | — | 84.5 | 83.0 | — | — | — | 47.0 | — | 79 | — | — | — |
| 33.0 | 45 | — | — | 84.0 | 82.5 | — | — | — | 46.2 | — | 79 | — | — | — |
| 32.5 | 44 | — | — | 83.5 | 82.0 | — | — | — | 45.5 | — | 78 | — | — | — |
| 32.0 | 43 | — | — | 82.5 | 81.5 | — | — | — | 44.8 | — | 77 | — | — | — |
| 31.5 | 42 | — | — | 82.0 | 81.0 | — | — | — | 44.0 | — | 76 | — | — | — |
| 31.0 | 41 | — | — | 81.5 | 80.5 | — | — | — | 43.4 | — | 75 | — | — | — |
| 30.8 | 40 | — | — | 81.0 | 79.5 | — | — | — | 43.0 | — | 74 | — | — | — |
| 30.5 | 39 | — | — | 80.0 | 79.0 | — | — | — | 42.1 | — | 74 | — | — | — |
| 30.0 | 38 | — | — | 79.5 | 78.5 | — | — | — | 41.5 | — | 73 | — | — | — |
| 29.5 | 37 | — | — | 79.0 | 78.0 | — | — | — | 40.7 | — | 72 | — | — | — |
| 29.0 | 36 | — | — | 78.5 | 77.5 | — | — | — | 40.0 | — | 71 | — | — | — |
| 28.5 | 35 | — | — | 78.0 | 77.0 | — | — | — | 39.3 | — | 71 | — | — | — |
| 28.0 | 34 | — | — | 77.0 | 76.5 | — | — | — | 38.5 | — | 70 | — | — | — |
| 27.8 | 33 | — | — | 76.5 | 75.5 | — | — | — | 37.9 | — | 69 | — | — | — |
| 27.5 | 32 | — | — | 76.0 | 75.0 | — | — | — | 37.5 | — | 68 | — | — | — |
| 27.0 | 31 | — | — | 75.5 | 74.5 | — | — | — | 36.6 | — | 68 | — | — | — |
| 26.5 | 30 | — | — | 75.0 | 74.0 | — | — | — | 36.0 | — | 67 | — | — | — |
| 26.0 | 29 | — | — | 74.0 | 73.5 | — | — | — | 35.2 | — | 66 | — | — | — |
| 25.5 | 28 | — | — | 73.5 | 73.0 | — | — | — | 34.5 | — | 66 | — | — | — |
| 25.0 | 27 | — | — | 73.0 | 72.5 | — | — | — | 33.8 | — | 65 | — | — | — |
| 24.5 | 26 | — | — | 72.5 | 72.0 | — | — | — | 33.1 | — | 65 | — | — | — |
| 24.2 | 25 | — | — | 72.0 | 71.0 | — | — | — | 32.4 | — | 64 | — | — | — |
| 24.0 | 24 | — | — | 71.0 | 70.5 | — | — | — | 32.0 | — | 64 | — | — | — |
| 23.5 | 23 | — | — | 70.5 | 70.0 | — | — | — | 31.1 | — | 63 | — | — | — |

## APPROXIMATE HARDNESS CONVERSIONS FOR STEEL

| Rockwell | | | | | | Rockwell Superficial | | | | Brinell | | Vickers | Shore | Approximate Tensile Strength (psi) |
| A 60 kg Brale | B 150 kg 1/16" Ball | C 150 kg Brale | D 100 kg Brale | E 100 kg 1/8" Ball | F 60 kg 1/16" Ball | 15-N 15 kg Brale | 30-N 30 kg Brale | 45-N 45 kg Brale | 30-T 30 kg 1/16" Ball | 3000 kg 10mm Ball Steel | 500 kg 10mm Ball Steel | 136° Diamond Pyramid | Sclero-scope | |
|---|---|---|---|---|---|---|---|---|---|---|---|---|---|---|
| 23.0 | 22 | — | — | 70.0 | 69.5 | — | — | — | 30.4 | — | 63 | — | — | — |
| 22.5 | 21 | — | — | 69.5 | 69.0 | — | — | — | 29.7 | — | 62 | — | — | — |
| 22.0 | 20 | — | — | 68.5 | 68.5 | — | — | — | 29.0 | — | 62 | — | — | — |
| 21.5 | 19 | — | — | 68.0 | 68.0 | — | — | — | 28.1 | — | 61 | — | — | — |
| 21.2 | 18 | — | — | 67.5 | 67.0 | — | — | — | 27.4 | — | 61 | — | — | — |
| 21.0 | 17 | — | — | 67.0 | 66.5 | — | — | — | 26.7 | — | 60 | — | — | — |
| 20.5 | 16 | — | — | 66.5 | 66.0 | — | — | — | 26.0 | — | 60 | — | — | — |
| 20.0 | 15 | — | — | 65.5 | 65.5 | — | — | — | 25.3 | — | 59 | — | — | — |
| — | 14 | — | — | 65.0 | 65.0 | — | — | — | 24.6 | — | 59 | — | — | — |
| — | 13 | — | — | 64.5 | 64.5 | — | — | — | 23.9 | — | 58 | — | — | — |
| — | 12 | — | — | 64.0 | 64.0 | — | — | — | 23.5 | — | 58 | — | — | — |
| — | 11 | — | — | 63.5 | 63.5 | — | — | — | 22.6 | — | 57 | — | — | — |
| — | 10 | — | — | 62.5 | 63.0 | — | — | — | 21.9 | — | 57 | — | — | — |
| — | 9 | — | — | 62.0 | 62.0 | — | — | — | 21.2 | — | 56 | — | — | — |
| — | 8 | — | — | 61.5 | 61.5 | — | — | — | 20.5 | — | 56 | — | — | — |
| — | 7 | — | — | 61.0 | 61.0 | — | — | — | 19.8 | — | 56 | — | — | — |
| — | 6 | — | — | 60.5 | 60.5 | — | — | — | 19.1 | — | 55 | — | — | — |
| — | 5 | — | — | 60.0 | 60.0 | — | — | — | 18.4 | — | 55 | — | — | — |
| — | 4 | — | — | 59.0 | 59.5 | — | — | — | 18.0 | — | 55 | — | — | — |
| — | 3 | — | — | 58.5 | 59.0 | — | — | — | 17.1 | — | 54 | — | — | — |
| — | 2 | — | — | 58.0 | 58.0 | — | — | — | 16.4 | — | 54 | — | — | — |
| — | 1 | — | — | 57.5 | 57.5 | — | — | — | 15.7 | — | 53 | — | — | — |
| — | 0 | — | — | 57.0 | 57.0 | — | — | — | 15.0 | — | 53 | — | — | — |

Hardness values in italics are beyond the normal range of that scale and are listed for reference only.

# TRIGONOMETRY FUNCTIONS

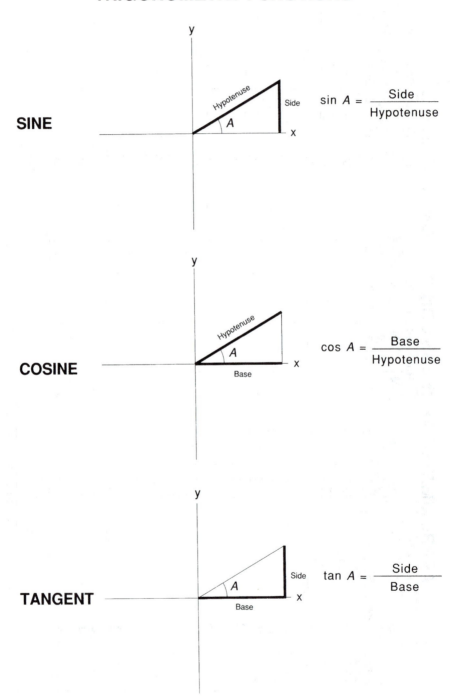

# RIGHT TRIANGLES

| Known Sides and Angles | Unknown Sides and Angles | | | Area |
|---|---|---|---|---|
| $a$ and $b$ | $c = \sqrt{a^2 + b^2}$ | $A = \text{arc tan } \dfrac{a}{b}$ | $B = \text{arc tan } \dfrac{b}{a}$ | $\dfrac{a \times b}{2}$ |
| $a$ and $c$ | $b = \sqrt{c^2 - a^2}$ | $A = \text{arc sin } \dfrac{a}{c}$ | $B = \text{arc cos } \dfrac{a}{c}$ | $\dfrac{a \times \sqrt{c^2 - a^2}}{2}$ |
| $b$ and $c$ | $a = \sqrt{c^2 - b^2}$ | $A = \text{arc cos } \dfrac{b}{c}$ | $B = \text{arc sin } \dfrac{b}{c}$ | $\dfrac{b \times \sqrt{c^2 - b^2}}{2}$ |
| $a$ and $\angle A$ | $b = \dfrac{a}{\tan A}$ | $c = \dfrac{a}{\sin A}$ | $B = 90° - A$ | $\dfrac{a^2}{2 \times \tan A}$ |
| $a$ and $\angle B$ | $b = a \times \tan B$ | $c = \dfrac{a}{\cos B}$ | $A = 90° - B$ | $\dfrac{a^2 \times \tan B}{2}$ |
| $b$ and $\angle A$ | $a = b \times \tan A$ | $c = \dfrac{b}{\cos A}$ | $B = 90° - A$ | $\dfrac{b^2 \times \tan A}{2}$ |
| $b$ and $\angle B$ | $a = \dfrac{b}{\tan B}$ | $c = \dfrac{b}{\sin B}$ | $A = 90° - B$ | $\dfrac{b^2}{2 \times \tan B}$ |
| $c$ and $\angle A$ | $a = c \times \sin A$ | $b = c \times \cos A$ | $B = 90° - A$ | $c^2 \times \sin A \times \cos A$ |
| $c$ and $\angle B$ | $a = c \times \cos B$ | $b = c \times \sin B$ | $A = 90° - B$ | $c^2 \times \sin B \times \cos B$ |

# OBLIQUE TRIANGLES

| Known Sides and Angles | Unknown Sides and Angles | | | Area |
|---|---|---|---|---|
| All three sides a, b, c | $A =$ arc cos $\dfrac{b^2 + c^2 - a^2}{2bc}$ | $B =$ arc sin $\dfrac{b \times \sin A}{a}$ | $C =$ $180° - A - B$ | $\dfrac{a \times b \times \sin C}{2}$ |
| Two sides and the angle between them a, b, ∠C | $c =$ $\sqrt{a^2 + b^2 - (2ab \times \cos C)}$ | $A =$ arc tan $\dfrac{a \times \sin C}{b - (a \times \cos C)}$ | $B =$ $180° - A - C$ | $\dfrac{a \times b \times \sin C}{2}$ |
| Two sides and the angle opposite one of the sides a, b, ∠A (∠B less than 90°) | $B =$ arc sin $\dfrac{b \times \sin A}{a}$ | $C =$ $180° - A - B$ | $c =$ $\dfrac{a \times \sin C}{\sin A}$ | $\dfrac{a \times b \times \sin C}{2}$ |
| Two sides and the angle opposite one of the sides a, b, ∠A (∠B greater than 90°) | $B =$ $180° -$ arc sin $\dfrac{b \times \sin A}{a}$ | $C =$ $180° - A - B$ | $c =$ $\dfrac{a \times \sin C}{\sin A}$ | $\dfrac{a \times b \times \sin C}{2}$ |
| One side and two angles a, ∠A, ∠B | $b =$ $\dfrac{a \times \sin B}{\sin A}$ | $C =$ $180° - A - B$ | $c =$ $\dfrac{a \times \sin C}{\sin A}$ | $\dfrac{a \times b \times \sin C}{2}$ |

# METRIC PREFIXES

| Prefix | Symbol | Multiplier |
|--------|--------|------------|
| tera | T | 1,000,000,000,000 |
| giga | G | 1,000,000,000 |
| mega | M | 1,000,000 |
| kilo | k | 1,000 |
| hecto | h | 100 |
| deka | da | 10 |
| | | 1 |
| deci | d | .1 |
| centi | c | .01 |
| milli | m | .001 |
| micro | μ | .000001 |
| nano | n | .000000001 |
| pico | p | .000000000001 |

# CONVERSION FACTORS

| Multiply | By | To Obtain |
|---|---|---|

## A

| | | |
|---|---|---|
| A | 1 | Amperes |
| Abcoulombs | $2.998 \times 10^{10}$ | Statcoulombs |
| Acre-feet | 43,560 | Cubic feet |
| Acre-feet | 1233.49 | Cubic meters |
| Acre-feet | 325,851 | Gallons |
| Acres | .4047 | Hectares |
| Acres | 43,560 | Square feet |
| Acres | 4,047 | Square meters |
| Acres | .001562 | Square miles |
| Acres | 4,840 | Square yards |
| Ampere-hours | 3,600 | Coulombs |
| Ampere-hours | .03731 | Faradays |
| Ampere-turns | 1.257 | Gilberts |
| Angstrom unit | $3.937 \times 10^{-9}$ | Inches |
| Angstrom unit | $10^{-10}$ | Meters |
| Ares | .02471 | Acres |
| Ares | 100 | Sq. meters |
| Astronomical units | $1.495 \times 10^{8}$ | Kilometers |
| Atmospheres | 1.01325 | Bars |
| Atmospheres | 76.0 | Cm. of mercury |
| Atmospheres | 33.90 | Feet of water |
| Atmospheres | 29.92 | Inches of mercury |
| Atmospheres | 10,333 | Kilograms/sq. meter |
| Atmospheres | 101.325 | KiloPascals |
| Atmospheres | 101,325 | Pascals |
| Atmospheres | 14.70 | Pounds/sq. inch |
| Atmospheres | 1.058 | Tons/sq. foot |

## B

| | | |
|---|---|---|
| Bags (cement) | 94 | Pounds |
| Barrels (cement) | 376 | Pounds |
| Barrels (dry) | 7056 | Cubic inches |
| Barrels (dry) | 105.0 | Quarts (dry) |
| Barrels (liquid) | 31.5 | Gallons |
| Barrels (oil) | 42 | Gallons |
| Bars | .9869 | Atmospheres |
| Bars | 1,000,000 | Dynes/sq. centimeter |
| Bars | 10,200 | Kilograms/sq. meter |
| Bars | 100 | KiloPascals |
| Bars | 750.0 | Millimeters of mercury |
| Bars | 100,000 | Pascals |
| Bars | 14.50 | Pounds/sq. inch |
| Baryls | 1 | Dynes/sq. centimeter |
| Board feet | 144 | Cubic inches |

| Multiply | By | To Obtain |
|----------|-----|-----------|
| Bolts (US Cloth) | 40 | Yards |
| B.T.U. | $1.055 \times 10^{10}$ | Ergs |
| B.T.U. | 778.2 | Foot-pounds |
| B.T.U. | 252.2 | Gram calories |
| B.T.U. | $3.930 \times 10^{-4}$ | Horsepower-hours |
| B.T.U. | 1,055 | Joules |
| B.T.U. | .2522 | Kilocalories |
| B.T.U. | 107.6 | Kilogram-meters |
| B.T.U. | $2.931 \times 10^{-4}$ | Kilowatt-hours |
| B.T.U. | 10.409 | Liter-Atmospheres |
| B.T.U./hour | .2162 | Foot-pounds/second |
| B.T.U./hour | .0700 | Gram-calories/second |
| B.T.U./hour | $3.930 \times 10^{-4}$ | Horsepower |
| B.T.U./hour | .2931 | Watts |
| B.T.U./minute | 12.97 | Foot-pounds/second |
| B.T.U./minute | .02358 | Horsepower |
| B.T.U./minute | .01758 | Kilowatts |
| B.T.U./minute | 17.58 | Watts |
| B.T.U./sq. foot/minute | .1221 | Watts/sq. inch |
| Buckets (British, Dry) | 18.18 | Liters |
| Bushels | 1.2445 | Cu. feet |
| Bushels | 2150.4 | Cu. inches |
| Bushels | .03524 | Cu. meters |
| Bushels | 35.24 | Liters |
| Bushels | 4 | Pecks |
| Bushels | 64 | Pints (dry) |
| Bushels | 32 | Quarts (dry) |

# C

| Multiply | By | To Obtain |
|----------|-----|-----------|
| C | 1 | Coulombs |
| Calories (gram) | .003966 | B.T.U |
| Calories (gram) | .001 | Calories (kilogram) |
| Calories (gram) | 41,840,000 | Ergs |
| Calories (gram) | .001 | Food calories |
| Calories (gram) | 3.0860 | Foot-pounds |
| Calories (gram) | $1.559 \times 10^{-6}$ | Horsepower-hours |
| Calories (gram) | .001 | Kilocalories |
| Calories (gram) | $1.162 \times 10^{-6}$ | Kilowatt-hours |
| Calories (kilogram) | 3.966 | B.T.U. |
| Calories (kilogram) | 1,000 | Calories (gram) |
| Calories (kilogram) | 1 | Food calories |
| Calories (kilogram) | 3,086 | Foot-pounds |
| Calories (kilogram) | .001559 | Horsepower-hours |
| Calories (kilogram) | 4,184 | Joules |
| Calories (kilogram) | 1 | Kilocalories |
| Calories (kilogram) | 426.8 | Kilogram-meters |
| Calories (kilogram) | .001162 | Kilowatt-hours |
| Candela/sq. centimeter | 3.142 | Lamberts |
| Candela/sq. inch | .4869 | Lamberts |

| Multiply | By | To Obtain |
|---|---|---|
| Candela/sq. meter | .0003142 | Lamberts |
| Carats | 3.086 | Grains |
| Centares (Centiares) | 1 | Square meters |
| Centigrams | .01 | Grams |
| Centiliters | .01 | Liters |
| Centimeter-dynes | .00102 | Centimeter-grams |
| Centimeter-dynes | $1.020 \times 10^{-8}$ | Meter-kilograms |
| Centimeter-dynes | $7.376 \times 10^{-8}$ | Pound-feet |
| Centimeter-grams | 980.7 | Centimeter-dynes |
| Centimeter-grams | $10^{-5}$ | Meter-kilograms |
| Centimeter-grams | $7.233 \times 10^{-5}$ | Pound-feet |
| Centimeters | .03281 | Feet |
| Centimeters | .3937 | Inches |
| Centimeters | .01 | Meters |
| Centimeters | 10 | Millimeters |
| Centimeters | 393.7 | Mils |
| Centimeters of mercury | .01316 | Atmospheres |
| Centimeters of mercury | .4461 | Feet of water |
| Centimeters of mercury | 136.0 | Kilograms/sq. meter |
| Centimeters of mercury | 27.85 | Pounds/sq. foot |
| Centimeters of mercury | .1934 | Pounds/sq. inch |
| Centimeters/second | 1.969 | Feet/minute |
| Centimeters/second | .03281 | Feet/second |
| Centimeters/second | .036 | Kilometers/hour |
| Centimeters/second | .6 | Meters/minute |
| Centimeters/second | .02237 | Miles/hour |
| Centimeters/second | $3.728 \times 10^{-4}$ | Miles/minute |
| Centimeters/sec./sec. | .03281 | Feet/sec./sec. |
| Centipoise | 1 | Millipascal-seconds |
| Centipoise | .001 | Pascal-seconds |
| Centistokes | $10^{-6}$ | Sq. meters/second |
| Centistokes | 1 | Sq. millimeters/second |
| Chains | 792 | Inches |
| Chains | 20.12 | Meters |
| Chains | 22 | Yards |
| Circular mils | $5.067 \times 10^{-6}$ | Sq. centimeters |
| Circular mils | $7.854 \times 10^{-7}$ | Sq. inches |
| Circular mils | .7854 | Sq. mils |
| Circumference | 6.2832 | Radians |
| Cord feet | 16 | Cu. feet |
| Cords | 8 | Cord feet |
| Coulombs | $1.036 \times 10^{-5}$ | Faradays |
| Coulombs | $2.998 \times 10^{9}$ | Statcoulombs |
| Coulombs/sq. centimeter | 64.52 | Coulombs/sq. inch |
| Coulombs/sq. inch | .1550 | Coulombs/sq. cm |
| Cubic centimeters | $3.531 \times 10^{-5}$ | Cubic feet |
| Cubic centimeters | .06102 | Cubic inches |
| Cubic centimeters | .000001 | Cubic meters |
| Cubic centimeters | $1.308 \times 10^{-6}$ | Cubic yards |
| Cubic centimeters | $2.642 \times 10^{-4}$ | Gallons |

| Multiply | By | To Obtain |
|---|---|---|
| Cubic centimeters | .001 | Liters |
| Cubic centimeters | .002113 | Pints (liquid) |
| Cubic centimeters | .001057 | Quarts (liquid) |
| Cubic feet | .8036 | Bushels (dry) |
| Cubic feet | 28,320 | Cubic centimeters |
| Cubic feet | 1728 | Cubic inches |
| Cubic feet | .02832 | Cubic meters |
| Cubic feet | .03704 | Cubic yards |
| Cubic feet | 7.48052 | Gallons |
| Cubic feet | 28.32 | Liters |
| Cubic feet | 59.84 | Pints (liquid) |
| Cubic feet | 62.43 | Pounds of water |
| Cubic feet | 29.92 | Quarts (liquid) |
| Cubic feet/minute | 472.0 | Cu. centimeters/second |
| Cubic feet/minute | .1247 | Gallons/second |
| Cubic feet/minute | .4720 | Liters/second |
| Cubic feet/minute | 62.43 | Pounds of water/minute |
| Cubic feet/second | .646317 | Millions of gallons/day |
| Cubic feet/second | 448.831 | Gallons/minute |
| Cubic inches | 16.39 | Cubic centimeters |
| Cubic inches | $5.787 \times 10^{-4}$ | Cubic feet |
| Cubic inches | $1.639 \times 10^{-5}$ | Cubic meters |
| Cubic inches | $2.143 \times 10^{-5}$ | Cubic yards |
| Cubic inches | .004329 | Gallons |
| Cubic inches | .01639 | Liters |
| Cubic inches | 106,100 | Mil-feet |
| Cubic inches | .03463 | Pints (liquid) |
| Cubic inches | .01732 | Quarts (liquid) |
| Cubic meters | 28.38 | Bushels (dry) |
| Cubic meters | 1,000,000 | Cubic centimeters |
| Cubic meters | 35.31 | Cubic feet |
| Cubic meters | 61,023 | Cubic inches |
| Cubic meters | 1.308 | Cubic yards |
| Cubic meters | 264.2 | Gallons |
| Cubic meters | 1000 | Liters |
| Cubic meters | 2113 | Pints (liquid) |
| Cubic meters | 1057 | Quarts (liquid) |
| Cubic yards | 764,600 | Cubic centimeters |
| Cubic yards | 27 | Cubic feet |
| Cubic yards | 46,656 | Cubic inches |
| Cubic yards | .7646 | Cubic meters |
| Cubic yards | 202.0 | Gallons |
| Cubic yards | 764.6 | Liters |
| Cubic yards | 1616 | Pints (liquid) |
| Cubic yards | 807.9 | Quarts (liquid) |
| Cubic yards/minute | .45 | Cubic feet/second |
| Cubic yards/minute | 3.367 | Gallons/second |
| Cubic yards/minute | 12.74 | Liters/second |
| Cycles/second | 1 | Hertz |

| Multiply | By | To Obtain |
|---|---|---|

# D

| | | |
|---|---|---|
| Days | 86,400 | Seconds |
| Decigrams | .1 | Grams |
| Deciliters | .1 | Liters |
| Decimeters | .1 | Meters |
| Degrees (angle) | 60 | Minutes |
| Degrees (angle) | .01111 | Quadrants |
| Degrees (angle) | .01745 | Radians |
| Degrees (angle) | 3600 | Seconds |
| Degrees (Celsius) | (°Cx9/5)+32 | Degrees (Fahrenheit) |
| Degrees (Celsius) | °C+273.15 | Degrees (Kelvin) |
| Degrees (Fahrenheit) | (°F-32)x5/9 | Degrees (Celsius) |
| Degrees (Fahrenheit) | °F+459.67 | Degrees (Rankine) |
| Degrees (Kelvin) | °K-273.15 | Degrees (Celsius) |
| Degrees (Rankine) | °R-459.67 | Degrees (Fahrenheit) |
| Degrees/second (angle) | .01745 | Radians/second |
| Degrees/second (angle) | .1667 | Revolutions/minute |
| Degrees/second (angle) | .002778 | Revolutions/second |
| Dekagrams | 10 | Grams |
| Dekaliters | 10 | Liters |
| Dekameters | 10 | Meters |
| Drams | 27.34375 | Grains |
| Drams | 1.771845 | Grams |
| Drams | .0625 | Ounces |
| Drams (troy) | .1371429 | Ounces |
| Drams (troy) | .125 | Ounces (troy) |
| Dyne-centimeters | 1 | Ergs |
| Dynes | 1 | Gram-cm./sec$^2$ |
| Dynes | $10^{-7}$ | Joules/centimeter |
| Dynes | $1.020 \times 10^{-6}$ | Kilograms |
| Dynes | $10^{-5}$ | Newtons |
| Dynes | $7.233 \times 10^{-5}$ | Poundals |
| Dynes | $2.248 \times 10^{-6}$ | Pounds |
| Dynes/centimeter | .01 | Ergs/sq.millimeter |
| Dynes/sq. centimeter | $9.869 \times 10^{-7}$ | Atmospheres |
| Dynes/sq. centimeter | $10^{-6}$ | Bars |

# E

| | | |
|---|---|---|
| Em (Pica) | .4233 | Centimeters |
| Em (Pica) | .167 | Inches |
| Ergs | $9.478 \times 10^{-11}$ | B.T.U. |
| Ergs | $2.390 \times 10^{-8}$ | Calories (gram) |
| Ergs | 1 | Dyne-centimeters |
| Ergs | $7.376 \times 10^{-8}$ | Foot-pounds |
| Ergs | .00102 | Gram-centimeters |
| Ergs | $3.725 \times 10^{-14}$ | Horsepower-hours |
| Ergs | $10^{-7}$ | Joules |
| Ergs | $2.390 \times 10^{-11}$ | Kilocalories |

| Multiply | By | To Obtain |
|---|---|---|
| Ergs | $1.020 \times 10^{-8}$ | Kilogram-meters |
| Ergs | $2.778 \times 10^{-14}$ | Kilowatt-hrs. |
| Ergs | $2.778 \times 10^{-11}$ | Watt-hours |
| Ergs/second | $5.688 \times 10^{-9}$ | B.T.U./minute |
| Ergs/second | 1 | Dyne-cm./sec. |
| Ergs/second | $4.427 \times 10^{-6}$ | Foot-pounds/minute |
| Ergs/second | $7.3756 \times 10^{-8}$ | Foot-pounds/second |
| Ergs/second | $1.341 \times 10^{-10}$ | Horsepower |
| Ergs/second | $1.433 \times 10^{-9}$ | Kilocalories/minute |

# F

| | | |
|---|---|---|
| F | 1 | Farads |
| Faradays | 26.80 | Ampere-hours |
| Faradays | 96,490 | Coulombs |
| Faradays/sec. | 96,490 | Amperes |
| Farads | 1,000,000 | Microfarads |
| Fathoms | 1.8288 | Meters |
| Fathoms | 6 | Feet |
| Feet | 30.48 | Centimeters |
| Feet | 12 | Inches |
| Feet | $3.048 \times 10^{-4}$ | Kilometers |
| Feet | .3048 | Meters |
| Feet | $1.894 \times 10^{-4}$ | Miles |
| Feet | $1.645 \times 10^{-4}$ | Miles (nautical) |
| Feet | 304.8 | Millimeters |
| Feet | .3333 | Yards |
| Feet of water | .02950 | Atmospheres |
| Feet of water | .8826 | Inches of mercury |
| Feet of water | .03048 | Kilograms/sq. cm. |
| Feet of water | 304.8 | Kilograms/sq. meter |
| Feet of water | 62.43 | Pounds/sq. foot |
| Feet of water | .4335 | Pounds/sq. inch |
| Feet/minute | .5080 | Centimeters/second |
| Feet/minute | .01667 | Feet/second |
| Feet/minute | .01829 | Kilometers/hour |
| Feet/minute | .3048 | Meters/minute |
| Feet/minute | .01136 | Miles/hour |
| Feet/second | 30.48 | Centimeters/second |
| Feet/second | 1.097 | Kilometers/hour |
| Feet/second | .5921 | Knots |
| Feet/second | 18.29 | Meters/minute |
| Feet/second | .6818 | Miles/hour |
| Feet/second | .01136 | Miles/minute |
| Feet/sec./sec. | 30.48 | Centimeters/sec./sec. |
| Feet/sec./sec. | .3048 | Meters/sec./sec. |
| Feet/sec./sec. | 1.097 | Kilometers/hr./sec. |
| Feet/sec./sec. | .6818 | Miles/hour/sec. |
| Foot candles | 10.764 | Lumens/sq. meter |
| Footlamberts | 3.426 | Candela/square meter |

| Multiply | By | To Obtain |
| --- | --- | --- |
| Foot-pounds | .001285 | B.T.U. |
| Foot-pounds | 13,560,000 | Ergs |
| Foot-pounds | .3240 | Gram calories |
| Foot-pounds | $5.051 \times 10^{-7}$ | Horsepower-hours |
| Foot-pounds | 1.356 | Joules |
| Foot-pounds | $3.240 \times 10^{-4}$ | Kilocalories |
| Foot-pounds | .1383 | Kilogram-meters |
| Foot-pounds | $3.766 \times 10^{-7}$ | Kilowatt-hours |
| Foot-pounds | 1.356 | Newton-meters |
| Foot-pounds/minute | .001285 | B.T.U./minute |
| Foot-pounds/minute | .01667 | Foot-pounds/sec. |
| Foot-pounds/minute | $3.030 \times 10^{-5}$ | Horsepower |
| Foot-pounds/minute | $3.240 \times 10^{-4}$ | Kilocalories/minute |
| Foot-pounds/minute | $2.260 \times 10^{-5}$ | Kilowatts |
| Foot-pounds/second | .07710 | B.T.U./minute |
| Foot-pounds/second | 4.6262 | B.T.U./hour |
| Foot-pounds/second | .001818 | Horsepower |
| Foot-pounds/second | .01944 | Kilocalories/minute |
| Foot-pounds/second | .001356 | Kilowatts |
| Furlongs | 660 | Feet |
| Furlongs | .125 | Miles |
| Furlongs | 40 | Rods |

# G

| Multiply | By | To Obtain |
| --- | --- | --- |
| g's (gravity) | 32.174 | feet/sec/sec |
| g's (gravity) | 9.806650 | meters/sec/sec |
| Gallons | 3785 | Cubic centimeters |
| Gallons | .1337 | Cubic feet |
| Gallons | 231.0 | Cubic inches |
| Gallons | .003785 | Cubic meters |
| Gallons | .004951 | Cubic yards |
| Gallons | .83267 | Gallons (Imperial) |
| Gallons | 3.785 | Liters |
| Gallons | 8 | Pints (liquid) |
| Gallons | 8.3453 | Pounds of water |
| Gallons | 4 | Quarts (liquid) |
| Gallons (Imperial) | 1.20095 | Gallons |
| Gallons/minute | 8.0208 | Cubic feet/hour |
| Gallons/minute | .002228 | Cubic feet/second |
| Gallons/minute | .06308 | Liters/sec. |
| Gallons/minute | 6.0086 | Tons of water/24 hrs. |
| Gausses | 6.452 | Lines/sq. inch |
| Gausses | $10^{-8}$ | Teslas |
| Gausses | $10^{-8}$ | Webers/sq. cm. |
| Gausses | $6.452 \times 10^{-8}$ | Webers/sq. in. |
| Gausses | $10^{-4}$ | Webers/sq. meter |
| Gilberts | .7958 | Ampere-turns |
| Gilberts/centimeter | .7958 | Ampere-turns/cm. |
| Gilberts/centimeter | 2.021 | Ampere-turns/inch |

| Multiply | By | To Obtain |
|---|---|---|
| Gilberts/centimeter | 79.58 | Ampere-turns/meter |
| Gills | .1183 | Liters |
| Gills | .25 | Pints (liquid) |
| Gills (British) | 142.07 | Cu. centimeters |
| Grade percentage | .9001 | Degrees |
| Grains | .0365714 | Drams |
| Grains | .06480 | Grams |
| Grains | .002083 | Ounces (troy) |
| Grains | .04167 | Pennyweights (troy) |
| Grains (troy) | 1 | Grains |
| Grains/gallon | 17.118 | Parts/million |
| Grains/gallon | 142.86 | Pounds/million gallons |
| Grains/Imp. gallon | 14.254 | Parts/million |
| Gram-centimeters | $9.295 \times 10^{-8}$ | B.T.U. |
| Gram-centimeters | 980.7 | Ergs |
| Gram-centimeters | $2.344 \times 10^{-8}$ | Kilocalories |
| Gram-centimeters | $10^{-5}$ | Kilogram-meters |
| Gram-centimeters/sec.$^2$ | 1 | Dynes |
| Grams | 980.7 | Dynes |
| Grams | 15.43 | Grains |
| Grams | .001 | Kilograms |
| Grams | 1000 | Milligrams |
| Grams | .03527 | Ounces |
| Grams | .03215 | Ounces (troy) |
| Grams | .07093 | Poundals |
| Grams | .002205 | Pounds |
| Grams/centimeter | .005600 | Pounds/inch |
| Grams/cu. centimeter | 62.43 | Pounds/cubic foot |
| Grams/cu. centimeter | .03613 | Pounds/cubic inch |
| Grams/liter | 58.417 | Grains/gallon |
| Grams/liter | 8.345 | Pounds/1000 gallons |
| Grams/liter | .062427 | Pounds/cubic foot |
| Grams/liter | 1000 | Parts/million |
| Grams/sq. centimeter | 2.0481 | Pounds/sq. foot |

# H

| Multiply | By | To Obtain |
|---|---|---|
| H | 1 | Henries |
| Hands | 4 | Inches |
| Hectares | 2.471 | Acres |
| Hectares | 10,000 | Square meters |
| Hectares | 107,600 | Square feet |
| Hectograms | 100 | Grams |
| Hectoliters | 100 | Liters |
| Hectometers | 100 | Meters |
| Hectowatts | 100 | Watts |
| Hertz | 1 | Cycles/second |
| Hogsheads | 8.42184 | Cubic feet |
| Hogsheads | 63 | Gallons |
| Hogsheads (British) | 10.114 | Cubic feet |

| Multiply | By | To Obtain |
|---|---|---|
| Horsepower | 42.41 | B.T.U./minute |
| Horsepower | 33,000 | Foot-pounds/min. |
| Horsepower | 550 | Foot-pounds/sec. |
| Horsepower | 1.014 | Horsepower (metric) |
| Horsepower | 10.69 | Kilocalories/minute |
| Horsepower | .7457 | Kilowatts |
| Horsepower | 745.7 | Watts |
| Horsepower (boiler) | 33,479 | B.T.U./hour |
| Horsepower (boiler) | 9.803 | Kilowatts |
| Horsepower (metric) | .9863 | Horsepower |
| Horsepower-hours | 2544 | B.T.U. |
| Horsepower-hours | 641,600 | Calories (gram) |
| Horsepower-hours | $2.685 \times 10^{13}$ | Ergs |
| Horsepower-hours | 1,980,000 | Foot-pounds |
| Horsepower-hours | 2,684,520 | Joules |
| Horsepower-hours | 641.6 | Kilocalories |
| Horsepower-hours | 273,700 | Kilogram-meters |
| Horsepower-hours | .7457 | Kilowatt-hours |
| Hundredweights | 1600 | Ounces |
| Hundredweights | 100 | Pounds |
| Hundredweights | .05 | Tons |
| Hundredweights | .0446429 | Tons (long) |
| Hundredweights | .0453592 | Tons (metric) |
| Hundredweights (long) | 112 | Pounds |
| Hundredweights (long) | .05 | Tons (long) |

# I

| | | |
|---|---|---|
| Inches | 2.54 | Centimeters |
| Inches | .0254 | Meters |
| Inches | 25.4 | Millimeters |
| Inches | 1,000 | Mils |
| Inches | .02778 | Yards |
| Inches of mercury | .03342 | Atmospheres |
| Inches of mercury | 1.133 | Feet of water |
| Inches of mercury | .03453 | Kilograms/sq. cm. |
| Inches of mercury | 345.3 | Kilograms/sq. meter |
| Inches of mercury | 70.73 | Pounds/sq. foot |
| Inches of mercury | .4912 | Pounds/sq. inch |
| Inches of water | .002458 | Atmospheres |
| Inches of water | .07355 | Inches of mercury |
| Inches of water | 25.40 | Kilograms/sq. meter |
| Inches of water | .5781 | Ounces/sq. inch |
| Inches of water | 5.202 | Pounds/sq. foot |
| Inches of water | .03613 | Pounds/sq. inch |

# J

| | | |
|---|---|---|
| J | 1 | Joules |
| Joules | $9.478 \times 10^{-4}$ | B.T.U. |
| Joules | 10,000,000 | Ergs |

| Multiply | By | To Obtain |
|---|---|---|
| Joules | .7376 | Foot-pounds |
| Joules | $3.725 \times 10^{-7}$ | Horsepower-hours |
| Joules | $2.390 \times 10^{-4}$ | Kilocalories |
| Joules | .1020 | Kilogram-meters |
| Joules | $2.778 \times 10^{-7}$ | Kilowatt-hours |
| Joules | 1 | Newton-meters |
| Joules | $2.778 \times 10^{-4}$ | Watt-hours |

# K

| Multiply | By | To Obtain |
|---|---|---|
| Kilocalories | 3.966 | B.T.U. |
| Kilocalories | 1 | Food calories |
| Kilocalories | 3086 | Foot-pounds |
| Kilocalories | 1000 | Gram calories |
| Kilocalories | .001559 | Horsepower-hours |
| Kilocalories | 4,184 | Joules |
| Kilocalories | 1 | Kilogram calories |
| Kilocalories | 426.9 | Kilogram-meters |
| Kilocalories | 4.184 | Kilojoules |
| Kilocalories | .001162 | Kilowatt-hours |
| Kilocalories/minute | 51.43 | Foot-pounds/second |
| Kilocalories/minute | .09351 | Horsepower |
| Kilocalories/minute | .06973 | Kilowatts |
| Kilogram-meters | 7.231 | Foot-pounds |
| Kilogram-meters/sec.$^2$ | 1 | Newtons |
| Kilograms | 1000 | Grams |
| Kilograms | 70.93 | Poundals |
| Kilograms | 2.205 | Pounds |
| Kilograms | .001102 | Tons |
| Kilograms/cu. meter | .001 | Grams/cu. centimeter |
| Kilograms/cu. meter | .06243 | Pounds/cu. foot |
| Kilograms/cu. meter | $3.613 \times 10^{-5}$ | Pounds/cu. inch |
| Kilograms/meter | .6720 | Pounds/foot |
| Kilograms/sq. cm. | .9678 | Atmospheres |
| Kilograms/sq. cm. | 980,665 | Dynes |
| Kilograms/sq. cm. | 32.81 | Feet of water |
| Kilograms/sq. cm. | 28.96 | Inches of mercury |
| Kilograms/sq. cm. | 2,048 | Pounds/sq. foot |
| Kilograms/sq. cm. | 14.22 | Pounds/sq. inch |
| Kilograms/sq. meter | $9.678 \times 10^{-5}$ | Atmospheres |
| Kilograms/sq. meter | $9.807 \times 10^{-5}$ | Bars |
| Kilograms/sq. meter | .003281 | Feet of water |
| Kilograms/sq. meter | .002896 | Inches of mercury |
| Kilograms/sq. meter | .2048 | Pounds/sq. foot |
| Kilograms/sq. meter | .001422 | Pounds./sq. inch |
| Kilograms/sq. millimeter | 1,000,000 | Kilograms/sq. meter |
| Kilolines | 1000 | Maxwells |
| Kiloliters | 1000 | Liters |
| Kilometers | 100,000 | Centimeters |
| Kilometers | 3281 | Feet |

| Multiply | By | To Obtain |
|---|---|---|
| Kilometers | 1000 | Meters |
| Kilometers | .6214 | Miles |
| Kilometers | 1094 | Yards |
| Kilometers/hour | 27.78 | Centimeters/second |
| Kilometers/hour | 54.68 | Feet/minute |
| Kilometers/hour | .9113 | Feet/second |
| Kilometers/hour | .5396 | Knots |
| Kilometers/hour | 16.67 | Meters/minute |
| Kilometers/hour | .6214 | Miles/hour |
| Kilometers/hr./sec. | 27.78 | Centimeters/sec./sec. |
| Kilometers/hr./sec. | .9113 | Feet/sec./sec. |
| Kilometers/hr./sec. | .2778 | Meters/sec./sec. |
| Kilometers/liter | 2.352 | Miles/gallon |
| KiloNewtons | 224.8 | Pounds |
| KiloPascals | .1450 | Pounds/sq. inch |
| Kilowatt-hours | 3412 | B.T.U. |
| Kilowatt-hours | $3.600 \times 10^{13}$ | Ergs |
| Kilowatt-hours | 2,655,000 | Foot-pounds |
| Kilowatt-hours | 1.341 | Horsepower-hours |
| Kilowatt-hours | 3,600,000 | Joules |
| Kilowatt-hours | 860.4 | Kilocalories |
| Kilowatt-hours | 367,100 | Kilogram-meters |
| Kilowatts | 56.87 | B.T.U./minute |
| Kilowatts | 44,250 | Foot-pounds/minute |
| Kilowatts | 737.6 | Foot-pounds/second |
| Kilowatts | 1.341 | Horsepower |
| Kilowatts | 14.34 | Kilocalories/minute |
| Kilowatts | 1000 | Watts |
| Knots | 1.689 | Feet/second |
| Knots | 1.8532 | Kilometers/hr. |
| Knots | 1.151 | Miles/hour |
| Knots | 1 | Nautical miles/hour |

# L

| | | |
|---|---|---|
| Lamberts | .3183 | Candela/sq. centimeter |
| Lamberts | 2.054 | Candela/sq. inch |
| Lamberts | 3183 | Candela/sq. meter |
| Leagues | 3 | Miles (nautical) |
| Light years | $9.4609 \times 10^{12}$ | Kilometers |
| Light years | $5.9 \times 10^{12}$ | Miles |
| Lines/sq. centimeter | 1 | Gausses |
| Lines/sq. inch | .1550 | Gausses |
| Lines/sq. inch | $1.550 \times 10^{-9}$ | Webers/sq. centimeter |
| Lines/sq. inch | $10^{-8}$ | Webers/sq. inch |
| Links (engineer's) | 12 | Inches |
| Links (surveyor's) | 7.92 | Inches |
| Liters | 1000 | Cubic centimeters |
| Liters | .03531 | Cubic feet |
| Liters | 61.02 | Cubic inches |

| Multiply | By | To Obtain |
|---|---|---|
| Liters | .001 | Cubic meters |
| Liters | .001308 | Cubic yards |
| Liters | .2642 | Gallons |
| Liters | 1.000 | Kilograms of water |
| Liters | 2.113 | Pints (liquid) |
| Liters | 1.057 | Quarts (liquid) |
| Liters/minute | $5.886 \times 10^{-4}$ | Cu. feet/second |
| Liters/minute | .004403 | Gallons/second |
| Lumens | .07958 | Spherical candle power |
| Lumens | .001496 | Watts |
| Lumens/sq. foot | 1 | Foot-candles |
| Lumens/sq. foot | 10.76 | Lumens/sq. meter |
| Lux | .0929 | Foot-candles |

# M

| Multiply | By | To Obtain |
|---|---|---|
| Maxwells | .001 | Kilolines |
| Maxwells | $10^{-8}$ | Webers |
| Megajoules | .2778 | Kilowatt-hours |
| Megalines | 1,000,000 | Maxwells |
| MegaPascals | 145.0 | Pounds/sq. inch |
| Meter-kilograms | 98,070,000 | Centimeter-dynes |
| Meter-kilograms | 100,000 | Centimeter-grams |
| Meter-kilograms | 7.233 | Pound-feet |
| Meters | 100 | Centimeters |
| Meters | 3.281 | Feet |
| Meters | 39.37 | Inches |
| Meters | .001 | Kilometers |
| Meters | $6.214 \times 10^{-4}$ | Miles |
| Meters | $5.396 \times 10^{-4}$ | Miles (nautical) |
| Meters | 1000 | Millimeters |
| Meters | 1.094 | Yards |
| Meters/minute | 1.667 | Centimeters/second |
| Meters/minute | 3.281 | Feet/minute |
| Meters/minute | .05468 | Feet/second |
| Meters/minute | .06 | Kilometers/hour |
| Meters/minute | .03238 | Knots |
| Meters/minute | .03728 | Miles/hour |
| Meters/second | 196.8 | Feet/minute |
| Meters/second | 3.281 | Feet/second |
| Meters/second | 3.6 | Kilometers/hour |
| Meters/second | .06 | Kilometers/minute |
| Meters/second | 2.237 | Miles/hour |
| Meters/second | .03728 | Miles/minute |
| Meters/sec./sec. | 3.281 | Feet/sec./sec. |
| Meters/sec./sec. | 3.6 | Kilometers/hr./sec. |
| Meters/sec./sec. | 2.237 | Miles/hour/sec. |
| Mhos | 1 | Siemens |
| Microfarads | $10^{-6}$ | Farads |
| Microns | $10^{-6}$ | Meters |

| Multiply | By | To Obtain |
|---|---|---|
| Mil-feet | $9.425 \times 10^{-6}$ | Cu. inches |
| Miles | 160,900 | Centimeters |
| Miles | 5280 | Feet |
| Miles | 63,360 | Inches |
| Miles | 1.609 | Kilometers |
| Miles | 1,609 | Meters |
| Miles | .8684 | Miles (nautical) |
| Miles | 1760 | Yards |
| Miles (nautical) | 6,080.27 | Feet |
| Miles (nautical) | 1.853 | Kilometers |
| Miles (nautical) | 1,853 | Meters |
| Miles (nautical) | 1.1516 | Miles |
| Miles (nautical) | 2,027 | Yards |
| Miles/gallon | .4251 | Kilometers/liter |
| Miles/hour | 44.70 | Centimeters/second |
| Miles/hour | 88 | Feet/minute |
| Miles/hour | 1.467 | Feet/second |
| Miles/hour | 1.609 | Kilometers/hour |
| Miles/hour | .8684 | Knots |
| Miles/hour | 26.82 | Meters/minute |
| Miles/hour | .1667 | Miles/minute |
| Miles/hour/sec. | 44.70 | Centimeters/sec./sec. |
| Miles/hour/sec. | 1.467 | Feet/sec./sec. |
| Miles/hour/sec. | 1.609 | Kilometers/hr./sec. |
| Miles/hour/sec. | .4470 | Meters/sec./sec. |
| Miles/minute | 2682 | Centimeters/second |
| Miles/minute | 88 | Feet/second |
| Miles/minute | 1.609 | Kilometers/minute |
| Miles/minute | 60 | Miles/hour |
| Milliers | 1000 | Kilograms |
| Milligrams | .01543 | Grains |
| Milligrams | .001 | Grams |
| Milligrams/liter | 1 | Parts/million |
| Milliliters | .001 | Liters |
| Millimeters | .1 | Centimeters |
| Millimeters | .03937 | Inches |
| Millimeters | .003281 | Feet |
| Millimeters | 39.37 | Mils |
| Millimeters of mercury | .001316 | Atmospheres |
| Millimeters of mercury | .001333 | Bars |
| Millimeters of mercury | .04461 | Feet of water |
| Millimeters of mercury | 13.60 | Kilograms/sq. meter |
| Millimeters of mercury | 133.3 | Pascals |
| Millimeters of mercury | .01934 | Pounds/sq. inch |
| Millimicrons | $10^{-9}$ | Meters |
| Million gallons/day | 1.54723 | Cu. feet/second |
| Mils | .00254 | Centimeters |
| Mils | .001 | Inches |
| Miner's inches | 1.5 | Cu. feet/minute |

| Multiply | By | To Obtain |
|----------|-----|-----------|
| Minims | .061612 | Cu. centimeters |
| Minims (British) | .059192 | Cu. centimeters |
| Minutes (angle) | .01667 | Degrees |
| Minutes (angle) | $1.852 \times 10^{-4}$ | Quadrants |
| Minutes (angle) | $2.909 \times 10^{-4}$ | Radians |
| Minutes (angle) | 60 | Seconds |

# N

| Multiply | By | To Obtain |
|----------|-----|-----------|
| N | 1 | Newtons |
| Nepers | 8.686 | Decibels |
| Newton-meters | .7376 | Foot-pounds |
| Newton-meters | 1 | Joules |
| Newtons | 100,000 | Dynes |
| Newtons | .1020 | Kilograms |
| Newtons | 1 | Kilogram-meters/$\text{sec}^{2}$ |
| Newtons | .2248 | Pounds |
| Newtons/sq. meter | 1 | Pascals |

# O

| Multiply | By | To Obtain |
|----------|-----|-----------|
| Oersteds | 79.58 | Amperes/meter |
| Ounces | 16 | Drams |
| Ounces | 437.5 | Grains |
| Ounces | 28.349527 | Grams |
| Ounces | .9115 | Ounces (troy) |
| Ounces | .0625 | Pounds |
| Ounces | $2.790 \times 10^{-5}$ | Tons (long) |
| Ounces | $2.835 \times 10^{-5}$ | Tons (metric) |
| Ounces (fluid) | 1.805 | Cubic inches |
| Ounces (fluid) | .02957 | Liters |
| Ounces (troy) | 480 | Grains |
| Ounces (troy) | 31.103481 | Grams |
| Ounces (troy) | 1.09714 | Ounces |
| Ounces (troy) | 20 | Pennyweights (troy) |
| Ounces (troy) | .08333 | Pounds (troy) |
| Ounces/sq. inch | .0625 | Pounds/sq. inch |

# P

| Multiply | By | To Obtain |
|----------|-----|-----------|
| Pa | 1 | Pascals |
| Parts/million | .0584 | Grains/gallon |
| Parts/million | .07016 | Grains/Imp. gallon |
| Parts/million | 8.345 | Pounds/million gallons |
| Pascals | $9.869 \times 10^{-6}$ | Atmospheres |
| Pascals | .00001 | Bars |
| Pascals | 1 | Newtons/sq. meter |
| Pascals | $1.450 \times 10^{-4}$ | Pounds/sq. inch |
| Pecks | .25 | Bushels |
| Pecks | 537.605 | Cubic inches |
| Pecks | 8.810 | Liters |
| Pecks | 8 | Quarts (dry) |

| Multiply | By | To Obtain |
|---|---|---|
| Pennyweights (troy) | 24 | Grains |
| Pennyweights (troy) | 1.55517 | Grams |
| Pennyweights (troy) | .05 | Ounces (troy) |
| Pennyweights (troy) | .004167 | Pounds (troy) |
| Pi | 1 | 3.141592654 |
| Picas | .1660 | Inches |
| Pints (dry) | 33.60 | Cu. inches |
| Pints (liquid) | 473.2 | Cu. centimeters |
| Pints (liquid) | .01671 | Cu. feet |
| Pints (liquid) | 28.87 | Cu. inches |
| Pints (liquid) | $6.189 \times 10^{-4}$ | Cu. yards |
| Pints (liquid) | .125 | Gallons |
| Pints (liquid) | .4732 | Liters |
| Pints (liquid) | .5 | Quarts (liquid) |
| Points | .01384 | Inches |
| Poise | 1 | Grams/centimeter-sec. |
| Pound-feet | 13,560,000 | Centimeter-dynes |
| Pound-feet | 13,825 | Centimeter-grams |
| Pound-feet | .1383 | Meter-kilograms |
| Poundals | 13,826 | Dynes |
| Poundals | 14.10 | Grams |
| Poundals | .001383 | Joules/centimeter |
| Poundals | .1383 | Joules/meter |
| Poundals | .01410 | Kilograms |
| Poundals | .1383 | Newtons |
| Poundals | .03108 | Pounds |
| Pounds | 256 | Drams |
| Pounds | 444,823 | Dynes |
| Pounds | 7000 | Grains |
| Pounds | 453.59237 | Grams |
| Pounds | .04448 | Joules/centimeter |
| Pounds | .45359237 | Kilograms |
| Pounds | .004448 | KiloNewtons |
| Pounds | 4.448 | Newtons |
| Pounds | 16 | Ounces |
| Pounds | 14.5833 | Ounces (troy) |
| Pounds | 32.17 | Poundals |
| Pounds | 1.21528 | Pounds (troy) |
| Pounds | .0005 | Tons |
| Pounds (troy) | 5760 | Grains |
| Pounds (troy) | 373.24177 | Grams |
| Pounds (troy) | 13.1657 | Ounces |
| Pounds (troy) | 12 | Ounces (troy) |
| Pounds (troy) | 240 | Pennyweights (troy) |
| Pounds (troy) | .822857 | Pounds |
| Pounds (troy) | $4.1143 \times 10^{-4}$ | Tons |
| Pounds (troy) | $3.6735 \times 10^{-4}$ | Tons (long) |
| Pounds (troy) | $3.7324 \times 10^{-4}$ | Tons (metric) |
| Pounds of water | .01602 | Cubic feet |

| Multiply | By | To Obtain |
|---|---|---|
| Pounds of water | 27.68 | Cubic inches |
| Pounds of water | .1198 | Gallons |
| Pounds of water/min. | $2.670 \times 10^{-4}$ | Cu. feet/second |
| Pounds/cubic foot | .01602 | Grams/cu. centimeter |
| Pounds/cubic foot | 16.02 | Kilograms/cu. meter |
| Pounds/cubic foot | $5.787 \times 10^{-4}$ | Pounds/cu. inch |
| Pounds/cubic inch | 27.68 | Grams/cu. centimeter |
| Pounds/cubic inch | 27,680 | Kilograms/cu. meter |
| Pounds/cubic inch | 1728 | Pounds/cu. foot |
| Pounds/foot | 1.488 | Kilograms/meter |
| Pounds/inch | 178.6 | Grams/centimeter |
| Pounds/sq. foot | .01602 | Feet of water |
| Pounds/sq. foot | 4.883 | Kilograms/sq. meter |
| Pounds/sq. foot | .006945 | Pounds/sq. inch |
| Pounds/sq. inch | .06804 | Atmospheres |
| Pounds/sq. inch | .06895 | Bars |
| Pounds/sq. inch | 2.307 | Feet of water |
| Pounds/sq. inch | 2.036 | Inches of mercury |
| Pounds/sq. inch | 703.1 | Kilograms/sq. meter |
| Pounds/sq. inch | 6.897 | KiloPascals |
| Pounds/sq. inch | .006897 | MegaPascals |
| Pounds/sq. inch | 6897 | Pascals |

# Q

| Multiply | By | To Obtain |
|---|---|---|
| Quadrants (angle) | 90 | Degrees |
| Quadrants (angle) | 5400 | Minutes |
| Quadrants (angle) | 1.571 | Radians |
| Quarts (dry) | 67.20 | Cubic inches |
| Quarts (dry) | 1.164 | Quarts (liquid) |
| Quarts (liquid) | 946.4 | Cubic centimeters |
| Quarts (liquid) | .03342 | Cubic feet |
| Quarts (liquid) | 57.75 | Cubic inches |
| Quarts (liquid) | $9.464 \times 10^{-4}$ | Cubic meters |
| Quarts (liquid) | .25 | Gallons |
| Quarts (liquid) | .9463 | Liters |
| Quarts (liquid) | .8594 | Quarts (dry) |
| Quintals | 100 | Kilograms |
| Quintals | 220.5 | Pounds |
| Quintals (Argentinean) | 101.28 | Pounds |
| Quintals (Brazilian) | 129.54 | Pounds |
| Quintals (Peruvian) | 101.43 | Pounds |
| Quintals (Chilean) | 101.41 | Pounds |
| Quintals (Mexican) | 101.47 | Pounds |
| Quires | 25 | Sheets |

# R

| Multiply | By | To Obtain |
|---|---|---|
| Radians | 57.30 | Degrees |
| Radians | 3438 | Minutes |
| Radians | .6366 | Quadrants |

| Multiply | By | To Obtain |
|---|---|---|
| Radians/second | 57.30 | Degrees/second |
| Radians/second | 9.549 | Revolutions/minute |
| Radians/second | .1592 | Revolutions/second |
| Radians/sec./sec. | 573.0 | Revolutions/min./min. |
| Radians/sec./sec. | .1592 | Revolutions/sec./sec |
| Reams | 500 | Sheets |
| Revolutions | 360 | Degrees |
| Revolutions | 4 | Quadrants |
| Revolutions | 6.283 | Radians |
| Revolutions/minute | 6 | Degrees/second |
| Revolutions/minute | .1047 | Radians/second |
| Revolutions/minute | .01667 | Revolutions/second |
| Revolutions/min./min. | .001745 | Radians/sec./sec. |
| Revolutions/min./min. | $2.778 \times 10^{-4}$ | Revolutions/sec./sec. |
| Revolutions/second | 360 | Degrees/second |
| Revolutions/second | 6.283 | Radians/second |
| Revolutions/second | 60 | Revolutions/minute |
| Revolutions/sec./sec. | 6.283 | Radians/sec./sec. |
| Revolutions/sec./sec. | 3600 | Revolutions/min./min. |
| Rods | .25 | Chains |
| Rods | 16.5 | Feet |
| Rods | 5.029 | Meters |
| Rods | 5.5 | Yards |

# S

| | | |
|---|---|---|
| S | 1 | Siemens |
| Scruples | 20 | Grains |
| Seconds (angle) | $2.778 \times 10^{-4}$ | Degrees |
| Seconds (angle) | .01667 | Minutes |
| Seconds (angle) | $3.087 \times 10^{-6}$ | Quadrants |
| Seconds (angle) | $4.848 \times 10^{-6}$ | Radians |
| Siemens | 1 | Mhos |
| Slugs | 14.59 | Kilograms |
| Slugs | 32.17 | Pounds |
| Square centimeters | 197,300 | Circular mils |
| Square centimeters | .001076 | Square feet |
| Square centimeters | .1550 | Square inches |
| Square centimeters | $10^{-4}$ | Square meters |
| Square centimeters | 100 | Square millimeters |
| Square feet | $2.296 \times 10^{-5}$ | Acres |
| Square feet | 929.0 | Square centimeters |
| Square feet | 144 | Square inches |
| Square feet | .09290 | Square meters |
| Square feet | $3.587 \times 10^{-8}$ | Square miles |
| Square feet | .1111 | Square yards |
| Square inches | 1,273,000 | Circular mils |
| Square inches | 6.452 | Square centimeters |
| Square inches | .006944 | Square feet |
| Square inches | 645.2 | Square millimeters |

| Multiply | By | To Obtain |
|---|---|---|
| Square inches | 1,000,000 | Square mils |
| Square kilometers | 247.1 | Acres |
| Square kilometers | 10,760,000 | Square feet |
| Square kilometers | 1,000,000 | Square meters |
| Square kilometers | .3861 | Square miles |
| Square kilometers | 1,196,000 | Square yards |
| Square meters | $2.471 \times 10^{-4}$ | Acres |
| Square meters | 10.76 | Square feet |
| Square meters | $3.861 \times 10^{-7}$ | Square miles |
| Square meters | 1.196 | Square yards |
| Square miles | 640 | Acres |
| Square miles | 27,878,400 | Square feet |
| Square miles | 2.590 | Square kilometers |
| Square miles | 3,097,600 | Square yards |
| Square millimeters | 1,973 | Circular mils |
| Square millimeters | .01 | Square centimeters |
| Square millimeters | .001550 | Square inches |
| Square mils | 1.273 | Circular mils |
| Square yards | $2.066 \times 10^{-4}$ | Acres |
| Square yards | 9 | Square feet |
| Square yards | 1,296 | Square inches |
| Square yards | .8361 | Square meters |
| Square yards | $3.228 \times 10^{-7}$ | Square miles |
| Stokes | .0001 | Sq. meters/second |

# T

| | | |
|---|---|---|
| T | 1 | Teslas |
| Teslas | 1 | Webers/sq. meter |
| Therms | 105.5 | Megajoules |
| Tonne | 1 | Ton (metric) |
| Tons | 907.18486 | Kilograms |
| Tons | 32,000 | Ounces |
| Tons | 29166.66 | Ounces (troy) |
| Tons | 2000 | Pounds |
| Tons | 2430.56 | Pounds (troy) |
| Tons | .89286 | Tons (long) |
| Tons | .90718 | Tons (metric) |
| Tons (long) | 1016 | Kilograms |
| Tons (long) | 2240 | Pounds |
| Tons (long) | 1.12 | Tons |
| Tons (metric) | 1000 | Kilograms |
| Tons (metric) | 2205 | Pounds |
| Tons (register) | 100 | Cubic feet |
| Tons (refrigeration) | 3.517 | Kilowatts |
| Tons of water/24 hrs. | 1.3349 | Cu. feet/hour |
| Tons of water/24 hrs. | .16643 | Gallons/minute |
| Tons of water/24 hrs. | 83.333 | Pounds water/hr. |
| Torr | .001333 | Bars |
| Torr | 133.3 | Pascals |
| Torr | .01934 | Pounds/sq. inch |

| Multiply | By | To Obtain |
|----------|-----|-----------|

## V

| | | |
|----------|-----|-----------|
| V | 1 | Volts |

## W

| | | |
|----------|-----|-----------|
| W | 1 | Watts |
| Watt-hours | 3.412 | B.T.U. |
| Watt-hours | 860.4 | Calories (gram) |
| Watt-hours | $3.6 \times 10^{10}$ | Ergs |
| Watt-hours | 2655 | Foot-pounds |
| Watt-hours | .001341 | Horsepower-hours |
| Watt-hours | .8604 | Kilocalories |
| Watt-hours | 367.1 | Kilogram-meters |
| Watt-hours | .001 | Kilowatt-hours |
| Watts | 3.412 | B.T.U./hour |
| Watts | .05687 | B.T.U./minute |
| Watts | $10^7$ | Ergs/second |
| Watts | 44.25 | Foot-pounds/minute |
| Watts | .7376 | Foot-pounds/second |
| Watts | .001341 | Horsepower |
| Watts | .001360 | Horsepower (metric) |
| Watts | .01434 | Kilocalories/minute |
| Watts | .001 | Kilowatts |
| Wb | 1 | Webers |
| Webers | 100,000 | Kilolines |
| Webers | $10^8$ | Maxwells |
| Webers/sq. inch | $1.550 \times 10^7$ | Gausses |
| Webers/sq. inch | $10^8$ | Lines/sq. inch |
| Webers/sq. inch | .1550 | Webers/sq. centimeter |
| Webers/sq. inch | 1,550 | Webers/sq. meter |
| Webers/sq. meter | 10,000 | Gausses |
| Webers/sq. meter | 64,520 | Lines/sq. inch |
| Webers/sq. meter | 1 | Teslas |
| Webers/sq. meter | .0001 | Webers/sq. centimeter |
| Webers/sq. meter | $6.452 \times 10^{-4}$ | Webers/sq. inch |

## Y

| | | |
|----------|-----|-----------|
| Yards | 91.44 | Centimeters |
| Yards | 3 | Feet |
| Yards | 36 | Inches |
| Yards | $9.114 \times 10^{-4}$ | Kilometers |
| Yards | .9144 | Meters |
| Yards | $5.682 \times 10^{-4}$ | Miles |
| Yards | $4.934 \times 10^{-4}$ | Miles (nautical) |

# INDEX

# *Notes*

# *Notes*

_____
_____
_____
_____
_____
_____
_____
_____
_____
_____
_____
_____
_____
_____
_____
_____
_____
_____
_____
_____
_____
_____
_____
_____
_____
_____

# *Notes*

## Notes

# Notes

*Notes*

## *Notes*

# *Notes*

# *Notes*

## Notes